Cooperative Extension   NRAES–117

Workforce Management for Farms and Horticultural Businesses
Finding, Training, and Keeping Good Employees

*Proceedings from the
Workforce Management for Farms
and Horticultural Businesses
Conference
Camp Hill, Pennsylvania
January 13–15, 1999*

NRAES—Natural Resource, Agriculture, and Engineering Service
Cooperative Extension
152 Riley-Robb Hall
Ithaca, New York 14853-5701

NRAES, the Natural Resource, Agriculture, and Engineering Service (formerly the Northeast Regional Agricultural Engineering Service) is an official activity of fourteen land grant universities and the U.S. Department of Agriculture. The following are cooperating members:

University of Connecticut Storrs, Connecticut	Rutgers University New Brunswick, New Jersey
University of Delaware Newark, Delaware	Cornell University Ithaca, New York
University of the District of Columbia Washington, DC	The Pennsylvania State University University Park, Pennsylvania
University of Maine Orono, Maine	University of Rhode Island Kingston, Rhode Island
University of Maryland College Park, Maryland	University of Vermont Burlington, Vermont
University of Massachusetts Amherst, Massachusetts	Virginia Polytechnic Institute and State University Blacksburg, Virginia
University of New Hampshire Durham, New Hampshire	West Virginia University Morgantown, West Virginia

NRAES–117
January 1999
© 1999 by NRAES—Natural Resource, Agriculture, and Engineering Service
All rights reserved. Inquiries invited. (607) 255-7654

ISBN 0-935817-37-9

Library of Congress Cataloging-in-Publication Data

Workforce Management for Farms and Horticultural Businesses Conference
 (1999 : Camp Hill, Pa.)
  Workforce management for farms and horticultural businesses :
finding, training, and keeping good employees : proceedings from the
Workforce Management for Farms and Horticultural Businesses
Conference, Camp Hill, Pennsylvania, January 13–15, 1999.
    p.   cm. — (NRAES ; 117)
  Includes bibliographical references (p.).
  ISBN 0-935817-37-9 (pbk.)
   1. Farms—Personnel management—Congresses. 2. Horticultural
products industry—Personnel management—Congresses. I. Natural
Resource, Agriculture, and Engineering Service. Cooperative
Extension. II. Title. III. Series: NRAES (Series) ; 117.
S563.6.W67 1999   98-51670
630'.68'3—dc21   CIP

Requests to reprint part(s) of this publication should be sent to NRAES. In your request, please state which part(s) of the publication you would like to reprint and describe how you intend to use the reprinted material. Contact NRAES if you have any questions.

NRAES—Natural Resource, Agriculture, and Engineering Service
Cooperative Extension, 152 Riley-Robb Hall
Ithaca, New York 14853-5701
Phone: (607) 255-7654 • Fax: (607) 254-8770 • E-mail: NRAES@CORNELL.EDU
Web site: HTTP://NRAES.ORG

Table of Contents

How Much Are Your Employees Worth? 1
    Robert A. Milligan, Bernard L. Erven

Communicating the Mission to All Personnel 10
    Lisa A. Holden

Managing the Multicultural Workforce 14
    Walter C. Montross

Performance Feedback 19
    Don R. Rogers

Recruiting and Hiring Outstanding Staff 24
    Bernard L. Erven

Getting the Most from Your Employees 35
    James G. Beierlein

Leadership: Coaching to Develop People 40
    Robert A. Milligan

Hiring with and without a Contract 45
    Jennifer LaPorta Baker

Elements of an Employment Contract 53
    John C. Becker

Guest Workers in Agriculture: The H-2A Temporary Agricultural Worker Program 58
    Al French

Discrimination in the Workplace 65
    Michael D. Pipa

Farm Employment Rules and Regulations: What You Need to Know 72
    Al French

EPA Worker Protection Standard (CFR Title 40. Part 170) 92
    David M. Scott

Developing a Safety Training Program Involves a Commitment to Reduce Hazards and
Injuries 104
    Ellen L. Abend, Eric M. Hallman

Table of Contents

Speaker Biographies

Jennifer LaPorta Baker 127
John C. Becker 128
James G. Beierlein 129
Bernard L. Erven 130
Al French 131
Lisa A. Holden 132

Robert A. Milligan 133
Walter C. Montross 134
Michael P. Pipa 135
Don R. Rogers 136
David M. Scott 137

Conference Notes Pages 138

How Much Are Your Employees Worth?

Robert A. Milligan, Ph.D.
J. Thomas Clark Professor of Entrepreneurship and Personal Enterprise
Department of Agricultural, Resource, and Managerial Economics
Cornell University

Bernard L. Erven, Ph.D.
Professor and Extension Specialist
Department of Agricultural, Environmental, and Development Economics
The Ohio State University

— *Robert A. Milligan's speaker biography appears on page 133* —
— *Bernard L. Erven's speaker biography appears on page 130* —

Introduction to People Oriented Management

Think about the following questions:

1. If an emergency call came for you right now, which of the following would be more upsetting to you?
    a. Your largest barn, processing shed, biggest piece of equipment and/or your best 50 head of livestock were just lost in a fire.
    b. Your three best employees have left and will not be back.

2. Do you know of a business that is succeeding while its people are failing?

3. Do you accomplish your goals - profit and whatever other goals you have -- through your hard work, your genius, or through your people?

4. Do you have a paradigm about people that fits your mission and goals?

In this paper we introduce a paradigm or way of looking at your business that views business personnel as the crucial asset to the success of your business. We refer to this as people oriented management. There is a growing body of research supporting the hypothesis that businesses successfully employing people oriented management are more successful in meeting their business goals than businesses with production oriented management.

Pfeffer(1998) summarizes research supporting this hypothesis from company performance after initial public offerings (IPOs) on the stock market; analyses from the steel, apparel manufacturing, the automobile, semiconductor manufacturing, and the oil refining industries: and several examples from service industries. Pfeffer (1998) and others argue that the increased success in people oriented businesses results from:

1. People work harder when informed, involved and empowered.
2. People work smarter when encouraged to learn and apply what they learn.
3. People stay longer when their career goals are being satisfied.
4. People build an attachment to their current place of employment when they feel important, engaged and "socially" involved.
5. Managers are more effective when enjoying a steady stream of people successes.

Pfeffer and the authors have identified from their research studies, related research, and personal observation and experience eight dimensions that characterize businesses that produce profits and other success measures through people. As you read and study this list, assess how your business is performing in this dimension:

1. They provide employment security.
2. They hire selectively. They expend the time to insure a large applicant pool, clearly articulate the required skills and attributes, select based on skills and attributes required for the position, and focus selection on attributes not easily changed through training.
3. Their organizational structures are decentralized and they often utilize self-managed teams.
4. They have above average compensation programs that are based on organizational performance.
5. They invest heavily in employee training.
6. They have less status distinctions and barriers including dress, language, working arrangements, and wage differentials across levels.
7. They have extensive sharing of financial and performance information.
8. In family businesses the business and family components must be harmony.

After looking at these dimensions, you may be saying WOW! How can I do this with my business? How can any business do all of this? Most do not but some successful businesses have had amazing success. Our purpose is to encourage you to become enthusiastic, even passionate, about becoming people oriented. We realize most of you have already started this journey; we hope to accelerate your journey to a people oriented business.

Can YOU Learn to be a Better Human Resource Manager?

The simple and emphatic answer is YES!!!! Unfortunately not all managers are convinced. We often hear managers say: "If I could communicate like Joe, I could be a better manager." Or "I don't have charisma; I can't mange people." Many years ago there was a great debate about whether managers were made or born. That debate, however, is over. **The skills, attitudes and behaviors that managers need can be learned**. The emphasis is, however, on you the manager to learn and implement these skills, attitudes and behaviors. We recognize, as must you, that YES is an easy answer but its implementation is a difficult, long, often frustrating journey. If you have not learned until now, what reason is there to believe you will now start learning? Ask yourself why you haven't learned to be a good HRM manager

We categorize the skills, attitudes and behaviors that managers need into three categories. The first category includes principally attitudinal factors:

1. Trust
2. Empathy
3. Humility
4. A willingness to share the glory.

The importance of these skills is emphasized by a recognition that most of what a manager does involves interpersonal relationships. Machines can be hooked together with no consideration of trust, empathy, humility and shared glory. Hooking people together through interpersonal relations cannot be so mechanical nor so physical. Covey (1989) stresses the importance of interpersonal relationships by describing the concept of an emotional bank account (Figure1) where you make deposits and withdrawals in the emotional bank account or trust level much like cash into and out of a checking account.

The second category is interpersonal skills including:

1. Communication
2. Conflict management and conflict resolution
3. Leadership
4. Discipline

These skills are the focus of several of the workshops on the second day of this conference. These skills require and build on the attitudes acquired as the first category of skills, attitudes and behaviors to be learned. One can be quite adept in one or two or even three of the four skills and still have major difficulties in building and sustaining interpersonal relations. To illustrate, an excellent communicator and leader may not understand the difference between conflict situations and discipline situations. This person's lack of conflict resolution skills could easily lead to lots of talk about a problem situation but little progress toward its resolution. In some cases, communication skills (a let's just talk about it approach) do not substitute for conflict resolution skills.

Figure 1. Emotional Bank Account

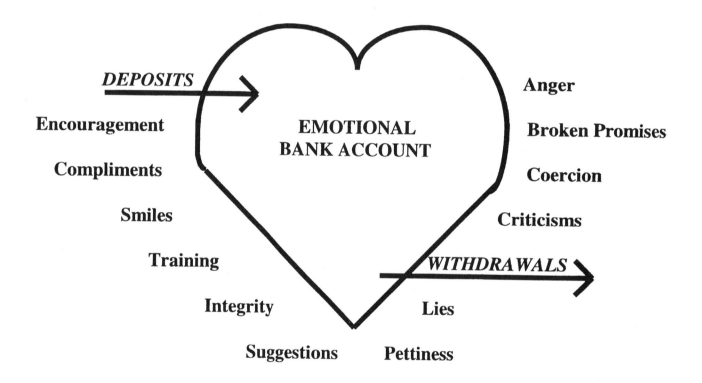

Source: Covey, Steven. R. 1991. **Principle-Centered Leadership**, Summit Books, NY.

The third category includes specific human resource management procedures including:
1. Determining labor and management needs, i.e., job analysis
2. Writing job descriptions
3. Recruiting, i.e., building a pool of applicants
4. Interviewing and using other selection tools
5. Orienting and training
6. Appraising performance
7. Developing and implementing compensation and benefit programs

A Framework for Learning About Human Resource Management

The starting point for learning about human resource management is an intuitive recognition and understanding of the message of modern management. Much has been written and many terms are used to describe modern management, but the authors believe the focus of modern management is:

**Developing and Utilizing ALL of the Capabilities of
ALL People in the Business or Organization**

The following are the three **Keys to Implementing Commitment-Oriented Management:**

A. Leaders and managers become developers and supporters. Begin with a recognition that people are assets to be developed-- not problems to be dealt with.

B. Commitment to a common vision/mission/purpose/goal.

C. Utilize empowerment and clear performance expectations rather than formal power to maintain business control.

Figure 2 contrast this focus on commitment-oriented management to the more traditional management paradigm based primarily on power.

Implementation of the first key requires that you recognize and internalize that employees are assets to be developed not problems to be dealt with. Employees are viewed as investments not costs. The interpersonal attitudes of trust and empathy and the interpersonal skills discussed above are key to successful implementation of this approach. Covey (1994) uses a catchy phrase of to live - to love - to learn - to leave a legacy to help us understand what is needed for personnel, including managers, to have high productivity and great job satisfaction. This occurs when their job provides:

FIGURE 2. LEADERSHIP/MANAGEMENT PARADIGMS: CONTROL VS. COMMITMENT

	Traditional Control Oriented	Deming Revolution Commitment-Oriented
Employee roles	Take orders Do your job	Ask questions Critical part of a system
Mechanism for achievement	Do the job "right"	Exceed expectations
Emphasis	Means/tasks	Ends/accomplishments
Development of people	Managers responsible for improvement Little need for training	Everyone responsible for improvements Training essential for continued success
Biggest fear	Upsetting the boss	Not meeting performance expectations
Structure	Bureaucratic Inherently adversarial	Flat Inherently team collegial
Employees' response	Demotivating	Motivating
Productivity	Less than the potential	Outstanding when successful

Developed by Dr. Robert A. Milligan, Cornell University, 305 Warren Hall, Ithaca, NY, 14850

To Live: Economic and physical needs (safety).

To Love: Social and emotional needs.

To Learn: Psychological need to use and develop talent.

To Leave a Legacy: Spiritual need to have meaning in what you do.

The second key focuses on commitment to a common vision or mission or we suggest organizational philosophy. This provides both the focus for employees being passionate about the business and the framework necessary for employees to become empowered to make decisions. The organizational philosophy (Rinke 1997) contains mission, vision and core values as described in the outline below:

Mission or Purpose:
- The "Why?"
- Individuals or organizations answer to "Why do we exist?"
- A broad statement of business scope and operations that distinguishes an organization from other similar organizations.
- What we get paid for.

Vision:
- The "What?"
- The picture of the future to be strived for.
- A motivational tool.
- Example: (Steven Jobs; Apple) To make a contribution to the world by making tools for the mind that advance mankind.
- One might be motivated in life to climb a ladder, but without vision, one might get to the wrong roof and thus have to start over.
- What we want to be when we grow up.

Core Values:
- Answer "How do we want to act, consistent with our mission and vision, along the path toward achieving our vision?"
- How individual or organization wants life to be day-to-day.
- Small number (3-5) and ranked.
- Example: Disney
    1. Safety
    2. Customers
    3. "The Show"
    4. Efficiency
- What's important around here?

The following are crucial points in understanding and implementing this commitment-oriented framework.

1. Understand where HRM fits in the bigger scheme of things in your business

2. Identify how you prefer to learn - experiences of others, workshops, reading materials, experimentation, videos and/or your own creative combinations
3. First things first - a five-year plan not a five-month plan.
4. Commitment to 360 degree evaluation and self-evaluation.

How to Implement People-oriented Management

The remainder of this conference is designed to help you implement people-oriented management. Keep the following points in mind as you embark on this exciting journey:

- Pave the way for change. Your senior employees may not welcome the important changes you see as essential. Some key members of your management team may also question the proposed changes. Engage these people in helping identify current human resource problems. Help them understand the changes you want to make and why you want to make them. Work to understand their reasons for resisting the changes you want to make. Use some combination of modification, negotiation, education, coercion and patience to win their support.

- Develop an abundance mentality. Emphasize how you can all be better off – a true win-win situation.

- Avoid over-emphasis on the problems with some individuals and concentrate instead on an enduring faith in people.

- Surround yourself with winners, i.e., hire great people. Your hiring practices and decisions have more long-run influence on your human resource success than any other single factor.

- Deal with your out-of-step people - some may have no future in your organization. Learn how to use progressive discipline so that these people must decide between leaving or getting on board with your new people-oriented management practices and expectations.

- Install a culture of continuous improvement to develop all the capabilities of all people.

- Unfreeze-change-refreeze

- Apply force-field analysis to increase success in implementing change.

Summary

We have only indirectly answered the question with which we started: "How much are your employees worth?" Now we can try a more direct answer.

Dollars and cents fail to measure the worth of your employees. Your words of praise for them only hint at their worth. Your knowing they are important even though you cannot express it is not enough. On the negative side, emotional outbursts about the problems people cause you measure your failings not their worthlessness.

Your employees' real worth is the value you place on your success, your goals and your accomplishing what you set out to accomplish when you hired them. Their worth and you are welded together. They and you together determine your joint worth. No more. No less.

Communicating the Mission to All Personnel

Lisa A. Holden
Assistant Professor
Department of Dairy and Animal Science
The Pennsylvania State University

— *Lisa A. Holden's speaker biography appears on page 132* —

Introduction

There is a story about a ship captain who gathered a crew to go on a great adventure. As they began their journey, the captain was busy assigning duties to everyone aboard. When one of the crew asked where they were headed, the captain replied, "You just worry about your job. Let me worry about our destination." One week into the journey, everything aboard ship was running smoothly when a severe storm rocked the ship and the captain was badly injured and unaware of his surroundings. When the storm subsided the captain was no better, and the crew argued about which direction to steer the ship. Unable to reach a decision and not knowing what the captain's plans were, the crew took the lifeboats in separate directions, headed for land.

Employees with no real sense of the business' mission may not "take to the lifeboats" like the ship's crew, but a good understanding of the direction of the business can improve working relationships and business performance. This paper will discuss some ways of communicating that mission to employees within the business and to customers, suppliers, and others outside the business.

The Mission Statement

Numerous books and papers have been written about what mission statements are, why mission statements and important, and how mission statements should be written. Some writers feel that mission statement should be short – short enough to fit on the back of a business card. Others feel that the mission statement should reflect the products and people of the business, however many words that takes. Some business mission statements are developed when the business is started. Some are developed and rarely modified, while other statements are reviewed an modified annually or on some other set time interval. There is no "right" or "wrong" way of developing a mission statement, and most agree that the mission statement should describe what the business is and what is does for its customers. There are also companies and consultants that help businesses develop mission statements.

The true value of the mission statement is not in having one printed on a business card or hung on a wall. The true value of having a mission statement is in using it as a means of communicating what is at the heart of a particular organization, so that every interaction among employees and between employees and owners and between employees and customers of that organization reflects the mission.

Why Communicate the Mission?

Most of us cannot imagine getting into a car with someone to travel from the East Coast to the West Coast if the driver doesn't have a map or a plan for how to make the trip. Yet, we often ask our employees and our customers to "go along with us" without that same road map that mission statement can provide. We just assume that they know that we are in business to make a profit, sell a high quality product, or provide good customer service.

Some businesses communicate their mission by posting it somewhere that employees, customers, and suppliers can read it. Walk into your local restaurant, bank, or bicycle shop and you will often see a mission statement hanging on the wall. Visit your local garden center, dairy farm, or apple orchard and you may also see a mission statement posted on a wall. While posting the mission statement for all to read is a nice place to begin; it is all too often the only way the mission statement becomes known. Thus, the true value of mission statements may never be realized.

One example of how a mission statement served as a yardstick against which all activities were measured was a well-known hotel that was owned by a large corporation. Employees in each division of the hotel were encouraged to develop a mission statement for their particular part of the business, then the hotel management used common elements in these division statements to develop the overall mission statement for the hotel. Employees were genuinely interested in the message conveyed in the overall mission since they had a part in its development. People in that organization provided services for guests, cleaned the facilities, and scheduled conferences with a shared sense of vision and purpose. Each division mission statement was posted below the overall hotel mission in various places where employees worked. Each employee had copies of both mission statements to that they could see how they contributed to the organization.

In some ways, it may be easier for larger firms to communicate mission to all the staff because the "corporate culture" of the business may be better understood. What do you think of when companies like General Motors, IBM or Microsoft are mentioned? Names like Apple Hill Farms, Hilltop Dairy, or Ridge Garden Center just don't bring to mind those same mental pictures. That may be why an understanding of the organizational mission statement may be even more important for small businesses. Your mission statement expresses your uniqueness in a highly competitive agricultural area. Not only does a clear understanding of the mission of the organization help employee performance, but it also helps to market your products and services to customers who could easily go to a competitor rather than buy from you.

Ideas for Making Mission Meaningful

Situation #1: A customer at a garden center wanted to return a plant that was purchased as a gift. A clerk at the garden center told the customer that he could only have credit towards another purchase. The customer argued that the person who sold him the plant said that it could be returned for a refund, if it was in good condition. A small tattered sign on the door of the shop read, "No returns accepted." Can a mission statement be helpful in this situation?

Situation #2: A morning milker for a dairy farm is frustrated by frequent interruptions in routine due to groups of school children visiting the farm. While other employees seem to enjoy the chance to answer a few questions, this milker feels uncomfortable since she was hired to milk, not talk to visitors. Can a mission statement be helpful in this situation?

Situation #3: A representative for a small agricultural supply company has just lost his biggest client due to repeated billing errors. The representative wants to offer an incentive to "win-back" this good customer, but is not certain how to handle making the offer. Can a mission statement be helpful in this situation?

Conclusions

A mission statement can be a useful tool in conveying that "mental picture" of the business that owners and managers have to all of the people involved in and served by that business. The elements of "corporate culture" are not clear in smaller farm businesses and a well understood mission statement could help employees to evaluate those day to day decisions using the same consistent yardstick. The real value of a mission statement is not capturing it on paper, but cultivating its use by the people who make it come to life each day.

References

Abrahams, Jeffrey. *The mission statement book: 301 corporate mission statements from America's top companies.* Berkeley, California: Ten Speed Press. 1995

Carver, John. *Creating a mission that makes a difference.* San Francisco, California: Jossey-Bass. 1997.

Haschak, Paul. *Corporate statements: the official missions, goals, principles, and philosophies of over 900 companies.* Jefferson, North Carolina: McFarland. 1998

Managing the Multicultural Workforce

Walter C. Montross
Certified Golf Course Superintendent
Westwood Country Club
Vienna, Virginia

— Walter C. Montross's speaker biography appears on page 134 —

     I forgot the basic rule when talking with professors...never indicate that you think you know something that they might find interesting or you may find yourself speaking in front of an audience. My topic "Managing the multi-cultural work force" is based on my experiences with a dynamic shift in my work force at Westwood Country Club. When I took the position in 1990 the basic makeup of the crew was white and black males. This staffing breakdown was very similar to other work forces I have employed over the years. Thus, my comfort level in managing this type of crew was strong based on familiarity. My managerial or leadership style leans toward "Authoritative". Based upon work done by David McClelland, he identifies this as managers who are "firm but fair". They tend to manage by providing clear instruction, soliciting some input (while leaving no doubt as to who is the boss), monitoring behavior and motivating with both discipline and rewards. They see influence as a key part of a manager's job. I mention this because things were about to change.

Also on my staff at the time was an older Hispanic man whose nationality was Bolivian. He spoke very poor English but had made his home in the United States for many years and had raised a family here. During my first couple of years at the club I was going through the normal personnel turnover. When I went to find replacement help I discovered that there were no people looking for this type of work at the level I could afford to pay. In Fairfax County where we are located the unemployment level has stayed between 1 ½ - 2 ½ % through the 1990's. As with most of you, my first step when looking for new help is to ask the crew if they know of anyone looking for work. Well, it just so happened that this same Hispanic man by the name of Rene Zenteno had a son looking for work. Little did I realize that things with regards to personnel were about to make a dramatic change. Without much thought by me a trend was started that eventually evolved into an almost complete change in the makeup of the staff. In 1990 the golf course staff at Westwood CC was 90% white, 10% black and the one Hispanic. In the past 8 years the crew has become 90% Hispanic and 10% white. The crew this past summer was made up of 15 Bolivians, 3 El Salvadorians and 3 white Americans. By tapping into a direct pipeline to the local Bolivian community, I have not had to advertise for help in over 6 years. I have had abundant help when all around me have struggled to find people. It is also important to note that the Club will not hire anyone that is not here legally.

The changes in the staff composition have been quite dramatic, but the changes that I have undergone have been equally challenging. I will be the first to admit that I was resistant to any change in my management style and I certainly wasn't willing to change my personality to suit anyone. Reflect back to the statement I made that I was an Authoritative Manager. I tell someone what to do and expect it done in a timely and efficient manner. I also expect that individuals will understand certain nuances and will do the little extra thing I have come to expect. The bottom line, this is the way I have always done it and I think it has worked fairly successfully. Now imagine my dealing with a new Hispanic employee, who has never seen a golf course and does not speak English. Even using an interpreter it is inevitable that things just won't work out. Try explaining things such as: a golf course has 18 holes, use this mower but not that one even though both mowers look the same but are set at different heights or simply trying to explain what just occurred to the green they were mowing when the individual wore soccer cleats to work.

Not only can the language barrier be a potential problem but many issues have come up based on cultural differences. The following is a list given to me many years ago by a ESL teacher at the local high school. This list addresses some cultural differences between Anglo-American and Hispanic-American peoples. Although there are exceptions to all rules I found this to hit pretty close to home. See if you see yourself and if you have had Hispanic employees see if you see certain traits.

Anglo-American	**Hispanic-American**
I am always on time. I live to work.	I take time to be happy. I work to live.
Plan for the future! You can shape your destiny.	Live now! Everything is in the hands of God.
I need my personal space. Don't come too close.	I am interested. I am fond of you. I get close and I touch you.
I am a busy and efficient person. I am polite.	I worry when you do not smile. When you smile I feel accepted.
Silence is golden.	Silence is rejection, sadness or anger.
I am impressed by your accomplishments. I identify with my job or career.	I am impressed by who you are. I identify with my boss or organization.
I am a parent: How is my child doing?	I am a parent: How is my child getting along?
I earn respect by my achievements.	I command respect by my position.
Common sense must never be lost.	We need passion to be truly alive.
I reflect on the consequences. I do not act on my feelings.	I am spontaneous. I act on impulse.
I value ideas over people.	I value people over ideas.
I celebrate victory. I try to overcome misfortune.	I admire stoicism. I try to except defeat.
I ask for help openly and directly.	I am embarrassed to admit I need help.
I acknowledge my errors, I apologize.	I do not blame myself. I blame fate.

Honesty is the best policy.	There are many sides to the truth.
First things first: Let's get down to business.	Let's get to know each other first.
I am pleased to meet you.	I am enchanted to meet you.

It should be obvious that my inability to speak Spanish, as well as my managerial style, had to lead to conflict. The lesson I offer today is pretty simple to say but pretty tough to execute. As they must adapt to a new country and a new job, I, too, had to be able to adapt.

I realized early on that the employees I was hiring worked hard, were on time and didn't complain about the menial nature of many of the jobs. Also they were willing to work for wages that did not attract anyone else. They enjoyed working outdoors and most importantly they enjoyed working with their fellow country men. Many of these fellows were only here in the United States for a while but some hoped to become permanent residents. Since these fellows have many options for other manual labor jobs, it certainly made it imperative that I take the steps to make it attractive for them to work for me. This also held the possibility of me finding people for long term employment. To accomplish this, I needed to change my thinking. I had to create a better system of communication and understanding. After all, the alternative is to not have the necessary help to get the job done. Some of the things we have done at the club to ease this communication barrier is to interpret everything both verbally and in writing. Employee handbooks, job descriptions, training manuals and signage are done in both English and Spanish to make sure there is full understanding. My assistant, mechanic and crew foreman all speak fluent Spanish. There is a full training program that takes place with any new hire to orient them to the nature of the golf course, as well as for operating any new piece of equipment.

Dealing with the language barrier is one thing but understanding the cultural differences is another matter entirely. Think back to the statement that Hispanic-American's interpret the lack of a smile as worrisome or silence means anger or sadness. I will be perfectly honest and say I seldom walk around with a smile on my face and I can be pretty quiet when focused on other matters. It is important that I let them know that this is my nature and that it shouldn't reflect on my feelings toward them. Periodically, I try and do cookouts and other activities that allow us to interact away from the job. I try to ask them about where they are from and how they feel about being in a different country. Personalizing things make them feel more a part of the bigger picture. The club gives all its employees their birthdays off. In the winter the club makes sure that families are invited to the Christmas parties. Each of these small things hopefully makes them feel welcome.

My resistance to change will find me enduring the ultimate sacrifice as I will be taking a course this winter to learn basic Spanish. I also hope to provide a class to my staff, that will teach English to those who want to learn.

It is inevitable that the manual labor force has and will continue to change. I read the other day that by the year 2050 it is estimated that the Hispanic population will be greater than any other within the United States. It only makes sense to utilize this available labor pool and most importantly to take the steps to make it work efficiently for you.

Performance Feedback

Don R. Rogers, AAC
Vice President/Consulting
First Pioneer Farm Credit, ACA
Enfield, Connecticut

— Don R. Rogers's speaker biography appears on page 136 —

Introduction

Getting the most out of your employees is critical to running a successful business in today's economy. Currently, scarce labor resources are forcing owners and supervisors to look at all conceivable ways to maximize human resources. There are many labor management tools that help orient, train, coach and get employees working together for a more profitable business. The various breakout sessions at this conference will illustrate many of these labor management tools.

This breakout session is specifically designed to talk about how to provide feedback to employees on how they are doing. Further, we will learn how to maximize the employees' contribution to improving key tasks and functions within the business. If a manager puts time into these two areas, it will pay tremendous dividends. Yet it is one of the most difficult tasks that a supervisor has to perform – communicating with their staff. Additional workshop material will be available as a supplemental proceeding.

Communications

Employees should have the opportunity to know how they are doing, where they stand and what they need to do a job that makes them feel good and help the business.

Employers like to have pride, and in return, have a level of comfort and security. All this starts with good communications.

Many businesses have some type of formal system for assessing job performance, whether you call it an annual review or a performance appraisal. This is very common in major businesses. It is usually described in the company's employee handbook. Here the process on how performance reviews are to be conducted are clearly explained and each manager should follow through with the same procedure.

In agriculture, reviews are seldom done on a formal basis. When a raise is given, perhaps the owner or supervisor will talk with the employee and tell them that he/she is appreciated. In fact, there is this syndrome of the "dreaded" review. Supervisors hate to give them; employees hate to get them. Reviews are usually done poorly and end up being a detriment instead of a positive event.

Why is Performance Feedback Useful?

Besides an employee knowing where they stand and having an opportunity to improve themselves, a performance review system has a lot of benefits. First of all, it becomes the motivating tool to have all employees working towards a common goal. If a business has a clear mission with measurable objectives, employees can see how their part fits. With some type of quantitative measure, employees can see how they are making a contribution towards the management of a business. Performance reviews also provide a chance to discuss what an employee does not know, what additional support you can provide and how you can get them to grow. Finding the right training and professional improvement opportunities allows the employee to grow and thus, the business grows.

Finally, a performance review system provides fairness and standardization on how you treat all employees. If an employee is not doing well, a formal review should clearly reflect their performance. Raises can be fairly given based on good performance standards or employees can be discharged for poor performance. It can eliminate bias and favoritism, especially when there are different supervisors in different departments but all within one system.

Why Performance Reviews are Dreaded?

Report cards are scary. Usually if you take a course and you have a good instructor who prepares you for the exams with a good lesson plan, you are comfortable about taking the test and knowing that you might get an A. If the teacher does a poor job, confuses you and if the final exam is a surprise and you get a D, this whole process is discouraging.

So when a supervisor does not do the right preparation, everyone loses. For instance, if a review is done only once a year with very little time involved and the supervisor does all the talking and dwells on all the areas that need improvement, an employee certainly wouldn't get excited about going through this. The employee can't wait to get it over. Thankfully, you only have to go through it once a year.

If the employee is surprised about the outcome in the review and if the supervisor's comments were bad, why did it take a whole year to find out? If the comments were great and you didn't know it, why did you have to wait the whole year to feel good? Performance reviews should not be a surprise. This is the fundamental rule of a successful system.

Formal Performance Review Process

If a business is going to install a <u>formal</u> review system then you need to put a lot of effort in it so that all employees and management benefit. Then there's a good chance it will be an effective tool. If it is tied to a salary increase or performance bonus, then it really needs to be done right. The employees should know what the expectation standards or measures would be. The employee should have all the necessary resources at their disposal so they could meet these expectations.

If it is an annual review, there should be progress checkpoints. If the employee needs training, the employer needs to allow this to happen and provide support. The formal performance review process needs to be conveyed to all employees at the time of hiring. All managers and supervisors who conduct performance reviews should be trained so that they can create an environment to make it a productive exercise.

How to Give a Performance Review

Giving a performance review is hard work. If you prepare and then prepare some more, you will feel comfortable. Just like giving a speech. If you have a good system and carry it out, then you and the employee would look forward to this time with their supervisor.

The rules of giving a performance review are simple.

- Set enough quality time away from disruptions. One hour minimum, hopefully longer.
- Use open-ended questions to bring employee into the discussion.
- Allow the employee to tell you how they think they are doing.
- You should start off with those things they do well.
- Highlight those areas that need improvement. At the conclusion, you and the employee agree on how you are going to follow through on these areas that need improvement.
- Always end on a positive note.
- If it is tied to a raise, you need to make sure that the raise and the performance review are consistent and fair.

Other Successful Techniques

A key employee should be involved in setting some of their own expectations. This will motivate them to achieve. A supervisor can modify them and direct them toward the business goal. A short progress session during the year is very useful. Encourage measuring progress quantitatively through a tracking system to measure speed, efficiency or quality of a job.

Self-assessment is probably the most powerful performance review system yet devised. Having an employee evaluate himself/herself is a great tool. Most employees are harder on themselves then their bosses. They can be very objective. When they see they need to improve, they are motivated to get the necessary training and resources to do a better job.

Getting Performance Feedback

A companion to this whole area of performance expectation is to remember that the important priority is for the business to be successful. If the business makes money, the employee has security and feels good about himself/herself. The company wins, the employee wins.

There are going to be tasks and areas of the business that are not performing well. Also, there is always new technology and new ways of getting the job done faster, better and cheaper. There are ways to be able to cut through the bottlenecks and even create a machine that does something better and faster. Companies that allow their employees to participate in correcting bottlenecks, solving problems and grabbing opportunities has a very powerful performance driven system. Great companies have managers that allow their employees the freedom to change. This technique of involving employees is so important that it can almost be the entire performance review system in itself.

In light of this, business owners need to open up their business to allow their employees to have a say in how they do their job and how the business changes. This needs a different mindset. It needs a new cultural philosophy. It needs an owner who is open-minded and not a control person.

Unfortunately, most agricultural businesses are driven by a family entrepreneur system with high risks (the owners sign on the debt and put their resources at risk) so there is a much more inherent need to control. Unfortunately, when the owners control things too much then their employees become only assistants. They are only there to hand the owner the tools or use equipment to get the job done. When owners are in such control they forget to delegate, they forget to allow others the freedom to get the job done. Therefore, they get bogged down with the task themselves. Most agriculture owners love to do physical activities. They are good at growing or visualizing building and equipment changes and are having fun doing it. So they take it on all by themselves. We see in many agriculture businesses, employees who are not asked for anything. So they just get their job done and go home.

Most businesses will fail or at least won't grow unless you can delegate responsibility downward. This allows the owner to become more of a manager and a visionary for the company. One of the most powerful labor management techniques is participative management. This simply means that employees have an opportunity to help direct, mold, develop and grow the company. The techniques of participative management are very simple. You create opportunities for the employees to help solve dilemmas. What we mean by this is simply to create a participative environment and let your employees go.

Here is a list of activities that will create participation.

1. Let employees hire, train and review their own staff.
2. Employees should be involved in standards to know how to improve things.
3. Have brainstorming meetings with those who are involved in a key task. For example, you want to speed the milking routine; you get some of the milkers together to offer suggestions. You want to improve the quality of the plants that you are growing. You get one of the growers, one from production and one from the sales group to review a particular crop and see how you can improve it.
4. Send employees to training with a purpose in mind. Have them report back what they learn and what they can change and then give them the tools to do it.
5. Encourage ideas, even if they are crazy, wild and impractical. Always thank them for their ideas and explain to them how you modified them a little bit to make them work.

In Summary

Giving performance feedback so that the employee knows how they are doing is hard work. It takes time and a system so that you are doing it fairly and effectively.

Utilizing the employees to look at opportunities to improve the business is a powerful human relation's tool. This requires a participative management approach and a leader or manager to be open minded and more of a coach then a dictator.

As the competition for labor becomes greater and the margins of your business harder to achieve, how you manage the human resource component of your business becomes the management challenge of the future.

References

- *The Employee Problem Solver* – Alexander Hamilton Institute, 70 Hill Top Road, Ramsey, NJ.
- *Personnel Management and Training* – Edward Harwell, Chain Store Publishing Corporation, New York.
- *The Motivational Manager*, Regan Communications, 212 W. Superior Street, Chicago, IL.
- *Writing Employee Handbook* – Thomas Maloney Cornell.
- *Effective Supervision Skills* – Dennis Murphy, Professional Training Associates, Round Rock, TX.
- *Increasing Employee Motivation* – Dennis Murphy, Professional Training Associates, Round Rock, TX.
- *The Performance Appraisal Sourcebook* - Baird/Beatty/Schneier, Human Resource Development Press, Amherst, MA

Recruiting and Hiring Outstanding Staff

Bernard L. Erven, Ph.D.
Professor and Extension Specialist
Department of Agricultural, Environmental, and Development Economics
The Ohio State University

— Bernard L. Erven's speaker biography appears on page 130 —

Farm managers face a major challenge in finding and keeping quality employees. Finding the quality employees is absolutely necessary if profitability, growth and excellence goals are to be reached. Family members and existing employees are unable to cover for missing employees for any extended period of time. Finding employees can become an urgent or even a critical problem threatening the future of the business.

No simple or even complex recipe guarantees hiring success. Luck is not the answer. The employer who seems lucky in always finding high quality people rarely is in fact lucky. Instead, such producers are depending on carefully made plans and a reputation as an excellent employer that has been patiently built. The answer lies in each farm employer developing a plan for filling positions.

The following eight step process for filling a position can be the foundation for a successful hiring plan.

Steps for Filling a Position

1. Determine the labor and management needs of the farm business that the new employee is expected to address
2. Develop a current job description based on the needs
3. Build a pool of applicants
4. Review applications and select those to be interviewed
5. Interview

6. Check references
7. Make a selection
8. Hire

Step 1 - Determine the labor and management needs of the farm business that the new employee is expected to address

What the business needs rather than what an applicant likes to do should guide the hiring process. An understanding of the goals for the farm business and its current and long-run constraints to progress will help in identifying desirable characteristics for employees. Goals and performance standards for areas with which the employee will have direct contact should be specifically addressed. This should happen before commencing the search for a new employee or starting a training program. This helps identify those specific things expected to be accomplished through hired farm workers in general and new employees in particular. For example, if an objective is to decrease repair costs, one alternative is to look for a person who has excellent mechanical skills. An alternative is to hire an inexperienced person who has a willingness and desire to master new skills. A follow up training program for such a person can result in a high quality employee.

Skipping this step and "hiring the first person in the driveway" is chancy. It means that a fit between what the farm needs and what the new employee brings to the job is left to chance.

Sometimes a farm manager is desperate for an employee. Taking time to think about the farm's needs seems unrealistic. The only way the farm manager can assure time for this step is to have backup labor already available and trained. The backup labor helps the farm through tight periods without forcing a hiring decision out of desperation.

Step 2 - Develop a Current Job Description Based on the Needs Identified in Step 1

Job descriptions help both the employer and employees by answering three questions: What does the jobholder do? How is it done? Under what conditions is it done? The job description has at least four parts:

1. Job title
2. A brief one or two sentence summary of the job
3. A listing of the major tasks involved in the job summarized under three to seven general headings, and
4. A listing of the knowledge, skills and abilities necessary to do the job.

Job descriptions may include other information such as the supervisor, who is supervised by the person in the position, pay range, required licenses or certificates, and location of the job.

Job descriptions are typically one page long. The brevity requires a terse, direct writing style. Simple words with single meanings should be used. Action verbs in the present tense should be used in defining the job duties, e.g., operates farm machinery including tractors, combines and trucks; enters data in computer, completes a performance evaluation at least annually for each employee supervised. The specifics of the job should be clear from the job description. The job title, job summary and description of duties should be completely consistent. To illustrate, the job title of herd manager is inconsistent with a list of job duties that includes only routine work tasks such as cleaning, feeding, moving, loading, and repairing.

Step 3 - Build a Pool of Applicants

Although there are many methods of getting job applicants, word of mouth and help wanted ads are likely to generate the most applicants. Word of mouth involves current employees, customers, neighbors, agribusiness contacts, veterinarians and others who come in contact with potential employees. Word of mouth is fast and low cost. However, it limits the scope of the job search because qualified applicants may not hear about the position. Current employees may quickly become advocates for relatives and friends.

The farm's reputation as an outstanding place to work is a powerful asset for generating a pool of applicants. Qualified people asking to fill out applications so they can be considered by the farmer the next time he has an opening is word of mouth working at its best.

Help wanted ads can be placed in newspapers and magazines known to be read by potential employees. Help wanted ads have the potential of expanding the applicant pool beyond the local community. The ads may increase the pool of applicants to the point that screening based on their application forms will be necessary.

Only well prepared and creative help wanted ads are likely to be effective in a tight job market. Other employers are working at least as hard as farmers to attract quality applicants. Following a seven-step process should result in an effective want ad:

1. Lead with a positive statement or job characteristic that attracts attention
2. Give the job title
3. Say something positive about the farm
4. Describe the job
5. Explain qualifications necessary for success in the position
6. Provide information on wages and benefits, as appropriate
7. Say how to apply for the job.

Two traditional and easily prepared help wanted ads illustrate what results when these seven steps are not followed:

> Wanted: Full-time worker for a vegetable farm. Call 888-9953.

> Experienced and reliable person needed for general farm work. Call 888-9953.

These ads have the advantage of costing little to run in a local newspaper. Their huge disadvantage is ineffectiveness. Traditional ads that do little to sell a job or the farm are unlikely to attract high quality applicants.

Following the seven steps for effective want ads results in nontraditional ads such as these:

> Looking for a change? Like farm work and animals but not long hours? We are a modern food producer specializing in pork looking for an ambitious individual to help care for our young livestock. You can start at 10:00 a.m. and be home with your family by 3:00. No experience needed - excellent training provided. Competitive wages and benefits. Weekend work optional. Submit your application at Sunrise Farms, Thursday-Saturday 4:00 p.m.- 6:00 p.m.

> Want to join a winning team? We are accepting applications for assistant farm manager of a modern, efficient vegetable farm. Responsibilities include training and supervising two full-time and four part-time employees, harvest equipment operation and maintenance, and other production related responsibilities. Previous farm experience, employee supervision and farm equipment operation desirable. Our excellent training program will help you succeed in this position. Attractive wage and fringe benefit package including health insurance. Call Kendra, Farm Manager at Valley Wide Farms, 613-888-9953.

The employer should be ready for telephone calls or visits from potential applicants. Each applicant should be asked to fill out an application form. Taking time to develop an application form or modify one used previously forces identification of important characteristics to look for in applicants. An application form provides a common base of information about all employees being considered. It also provides an important source of questions to be followed up on during the interview. The application form should include an agreement section signed by the applicant which gives permission to check references, makes clear that false information on the application form can be basis for dismissal and that the employment is at-will.

Step 4 - Review Applications and Select
Those to be Interviewed

Some applicants will be excluded from further consideration based on the application form. A pre-interview can also be used to help identify applicants to be invited for a formal interview. Having interested people visit the farm to fill out an application form can provide opportunity for a few general questions about experience and interest in the job. Promising candidates can be given a mini-tour of the farm providing opportunity for general conversion about farming, livestock, the swine industry, farm work and machinery. The objective of the pre-selection step is to reduce the applicant pool to the most promising candidates.

No fewer than three people should be left in the applicant pool. You may not be successful in hiring the best person in the pool of applicants. Interviewing may dramatically change the pre-ranking of applicants you have made. Also, some applicants will withdraw. Most important, the person hired should know that he or she is a winner having been selected over other qualified people.

Step 5 - Interview

The following questions need to be addressed in preparation for interviewing:

1. Who will be on the interview team?
2. How will time be divided between the formal interview and informal discussion including a visit of the farmstead?
3. What questions will be asked in the interview?
4. How will interview evaluations be recorded?
5. Where will the interview be conducted?

Preparing a list of questions before the interview is critical to interview success. Avoid questions that can be answered yes or no. Instead of yes/no type questions, use open-ended questions that encourage applicants to explain experiences, characteristics and ideas in their own words. The open-ended questions should be geared toward the following general areas: previous job accomplishments and achievements, non-job accomplishments and achievements, motivation and ambition, hobbies and use of leisure time, and "what if" questions. "What if" questions are based on practical real-world problem situations. The intent is to discover how the applicant would handle the real-world problem.

So that applicants can be compared on the same criteria, the basic list of questions asked should be the same for each applicant.

Do not ask questions about: race, color, religion, national origin, marital status, number and care of dependents, height, weight, education unrelated to the job, friends or relatives who have previously worked on your farm, age, arrest records, U.S. citizenship, disabilities, person to notify in case of emergency, sexual orientation, nonbusiness-related references, social clubs and organizations, and military experience in the armed forces of another country. A general guideline is to ask only about those things that are unquestionably related to the job and any applicant's ability to do the job.

It is possible to get necessary information without asking improper questions. It is legitimate to ask about availability for work on weekends and staying late during planting and harvest seasons. However, these questions should not be asked in terms of family responsibilities, children or religious practice. It is important to know if an applicant is a U.S. citizen or whether the applicant meets immigration law requirements. These questions can be asked without reference to national origin.

Once you have selected the interviewing team and planned the interview, you are ready to interview the people selected in Step 4. The interview can be divided into the following nine steps:

1. Relax the applicant and build rapport. (2-3 minutes)
2. Give the applicant a copy of the job description and describe the job in considerable detail. (3-5 minutes)
3. Determine the accuracy of the information on the application form. (4-7 minutes)
4. Ask a series of open-ended questions previously prepared. (10-15 minutes)
5. Encourage the applicant to ask questions. (2-5 minutes)
6. Summarize your farm's mission, objectives, and business philosophy. (2-4 minutes)
7. Summarize the opportunities provided to the person in the position. (2-4 minutes)
8. Encourage the applicant to ask questions. (2-10 minutes)
9. Close with information about plans for making a decision. (2-4 minutes)

The interview should typically take about 30 minutes. The times suggested in parentheses can be adjusted for shorter or longer interviews.

The content, importance and intent of each step should be thoroughly understood before beginning interviews.

1. Relax the applicant and build rapport. (2-3 minutes)

   Although this step should take no more than 2-3 minutes, it is important to all the steps that follow in the interview. The objective is to set the stage for a friendly and open exchange of information. Your smile and warm welcome are important. Possible discussion topics include the weather, any difficulty in finding the farm, a school attended by both interviewer and interviewee, or a friendly dog who enthusiastically greets all visitors. Confirming that the applicant has parked in the right place may be helpful. Maintain a casual and non-interview atmosphere during this step.

2. Give the applicant a copy of the job description and describe the job in considerable detail (3-5 minutes)

It is essential that the applicant understand the job you are filling. Do not depend on general terms like milking cows, taking care of pigs, driving tractor, waiting on customers and general farm work. The meanings of these terms vary substantially from farm to farm. Be specific about the duties and responsibilities.

3. Determine the accuracy of the information on the application form. (4-7 minutes)

Review the applicant's training directly required for performance of the job, job experience directly related to the position you are filling, and gaps of time between jobs. Pay particular attention to career progress from one job to the next and vague reasons for leaving previous positions.

4. Ask a series of open-ended questions previously prepared. (10-15 minutes)

Your careful preparation for the interview should be apparent to the applicant. Avoid groping for the next question to ask. Impress the applicant with your ability to ask questions that are fun to answer. You are conducting an interview not an interrogation. Embarrassing and stressing applicants is unnecessary.

5. Encourage the applicant to ask questions. (2-5 minutes)

Note that thus far in the interview, the applicant has been responding. At this point, the applicant is given explicit encouragement to ask questions. Be patient in allowing time for the applicant to get his or her questions formulated and asked. You should answer the applicant's questions in a straightforward manner. Note carefully the content of the questions, the insight shown, and follow up questions when pursuing a particular point. Pay careful attention to hints about the career and personal needs the applicant hopes to satisfy through the job.

6. Summarize your farm's mission, objectives, and business philosophy. (2-4 minutes)

This is a "selling" step. You want the applicant to have a positive impression of your business even if an offer will not be forthcoming. Take time to explain the uniqueness of your business, the importance of people in accomplishing your goals and your vision of the opportunities in agriculture and the swine industry in particular. Also explain the pride you have in former employees who have moved up in the industry.

7. Summarize the opportunities provided to the person in the position. (2-4 minutes)

You now turn from the general summary about the farm business to a specific summary about the position you are filling. This is also a "selling" step. It is appropriate to explain again the importance of the position to the success of your business, the opportunities there will be to learn the necessary skills for success, and the satisfaction that can be gained through the position.

8. Encourage the applicant to ask questions. (2-10 minutes)

   This second opportunity for the applicant to ask questions should be used to emphasize your desire to be open and an effective communicator. Show your caring attitude. The applicant may have thought of additional questions or now have the courage to ask questions that earlier seemed too daring. This second opportunity for the applicant to ask questions further encourages the applicant to interview you instead of just being interviewed by you.

9. Close with information about plans for making a decision. (2-4 minutes)

   Be specific about what happens next, when you will complete applicant interviews and when you plan to be in touch with the applicant. Be sure the applicant does not leave guessing about what the next step is.

   Be careful not to raise unrealistic expectations for the applicant. Simply express appreciation for the applicant's time, provide your name and telephone number, and welcome personal contacts with you should the applicant have any questions.

Interviewing is difficult. Knowing how to do it well makes it enjoyable. Some dos and don'ts can serve as reminders on how to improve your interviewing skills.

Do:
1. Make sure the applicant does most of the talking.
2. Make the interview fun for you and the applicant.
3. Listen!!!
4. Be attentive.
5. Concentrate on the interview and what is being said.
6. Show enthusiasm throughout the interview.
7. "Read" non-verbal messages.
8. Show appreciation for the person being interested in the position.
9. Show pride in your business, agriculture and the swine industry.
10. Stay in control of the interview.

Don't:
1. Project the answer you want from the applicant, e.g., "You do like pigs don't you."
2. Cut an interview short because the first ten minutes did not go well.
3. Let your note taking during the interview detract from the "flow" on the

interview.
4. Read questions to the applicant.
5. Let your facial expressions and other non-verbal responses show your dissatisfaction with the applicant's answers.
6. Add a series of follow up questions to explore "interesting" side issues.
7. Allow an aggressive applicant to ignore your questions and talk about things not on your agenda.
8. Go into the interviews with the intention of simply confirming that a pre-interview favorite is in fact the best candidate for the position.

Step 6 - Check References

References can confirm information gathered in steps 1-5 and provide additional information about those applicants who are still being given serious consideration. Some employers skip this step because of previous employers' reluctance to share any useful information out of fear of defamation charges.

Reference checks can still be productive. Personal visits or telephone conversations will be more productive than asking for written comments. Getting references from your personal acquaintances or from people well known in agricultural circles will be more productive than asking strangers. Asking about the most important contribution the employee has made is likely to be more helpful than asking if the reference knows of any reason you should not hire the person. A reference's tone of voice may communicate more than the words being said. Asking references provided by the applicant to suggest other people to contact can result in additional useful information.

Keep in mind that some references have reason to give less than candid information. Some employers may praise a problem employee in hopes that another employer will hire away the person. On the other hand, some employers may hint at problems in hopes of preventing you from making an offer to one of their outstanding employees.

Asking the same carefully prepared questions of each reference will be helpful. Using a structured form can greatly simplify recording information received from references.

Step 7 - Make a selection

Strive to be as objective as possible given the job description; knowledge, skills and abilities necessary to do the job; and the information available concerning each applicant. If no satisfactory applicant is found, start the process over rather than deciding to take a chance on a doubtful applicant.

Selection biases can easily creep into the selection process. Five potentially important selection biases are:

1. Stereotyping: Attributing certain characteristics to a particular group of people. "People who grew up on livestock farms like animals."

2. Halo effect: Regarding highly an individual who has characteristics you particularly like.

   "A person, like me, who drives a Chevy, loves country music and is a Chicago Cubs fan will be a good employee."

3. First Impressions: Judging prematurely based on appearance, handshake or voice. "He has a good firm handshake, a friendly smile, no earrings, and short hair. I knew before the interview started that he would be a good employee."

4. Contrast: Measuring against the last person interviewed. "After that last person we interviewed, I had begun to think we would never be able to find a good person."

5. Staleness: Discounting those interviewed early and favoring a person interviewed just before the selection is made.

Step 8 - Hire a person

Make an oral offer in person or by telephone to your first choice followed by a written offer that summarizes the key conditions of employment. In making the offer, emphasize that the applicant is the first choice among several qualified people. Show enthusiasm over the hope that this person will soon be joining your farm team.

The written employment agreement can be a letter of explanation or a form with blanks filled in as appropriate. Whatever the form, the agreement should include a description of the job, a statement that the employment is "at will," and explanation of compensation, benefits, work schedules and any other important details.

Summary

"This is all nice but... I don't have enough time to follow all these steps."

"This is all nice but... I don't know how to do all these things you say are necessary."

"This is all nice but... I only need the Ford of interviewing not the Mercedes you have described."

"This is all nice but... I don't know any farmer who pays this much attention to filling a position."

The recipe for farm success is complex. Animals, equipment, financing, land and buildings matter a great deal. People also matter. To a great extent, managers reach their goals through people. Getting things done through people requires competent employees. Mediocrity in filling positions can make a huge difference over time. To have competent employees, people who have the potential of being competent need to be hired. The question is: Do I maximize my chances of hiring the "right" people or do I leave my success to chance? Each farm employer answers this question directly or indirectly and then lives with the answer.

Getting the Most from Your Employees

James G. Beierlein, Ph.D.
Professor
Department of Agricultural Economics and Rural Sociology
The Pennsylvania State University

— James G. Beierlein's speaker biography appears on page 129 —

Introduction

One of the greatest challenges facing any business is the ability to remain competitive in a rapidly changing market. Business success today is increasingly being defined by the quality and quantity of a firm's human capital rather than by its financial or physical assets. Management's greatest test is its ability to define and find the highest quality human capital. Next they must build a work environment that allows human capital to grow, remain current, and be coordinated so that the firm can accomplish the objectives set forth in its business plan. After all, a firm can be no better than the people who work there.

The Role of a Good Business Plan

The job of getting the most out of your employees is made a good deal easier when managers have a firm grasp of the principles of human motivation and clear business objectives. Clear business objectives come from a well-written business plan that gives clear, concise statements that define the business's purpose (*"the what"*) and objective (*"the how"*). When the owners of the firm can tell their employees, customers, vendors, and others *what* the firm is going to do, and *how* they are doing to satisfy customers' needs better, faster, quicker, cheaper than anyone else the firm is ready to *start* building its business.

All too often the leaders of a business have a vision in their mind of what they want to do. Most managers keep it safely tucked away and rarely talk about it or even share it with others. Fewer yet write it down. After a while, all they remember is the answer. The questions of "the what" and "the how" are lost. The result is often the program of the week. The only question is whether you get a T-shirt or a coffee mug with the latest program. Adherence to basic business principles and continued focus on the satisfaction of customer needs are the thing that keeps businesses successful over the long haul. There are many examples of businesses that lost sight of these points and paid for their myopia with huge financial losses, management turmoil, and even bankruptcy. When your employees do not have a clear vision of what you expect of them chaos ensues. A well-written business plan and clear communication of its contents are prerequisites to business success.

A business plan is just a plan. The real value of any plan is in how it helps a business accomplish its purpose and objectives. A key ingredient in making this transition is getting all the employees to accept the firm's purpose and objective as worthwhile, and to take them on as personal goals. Moving toward these in lockstep is not necessary, but everyone should at least be moving in the same direction. People will do this if they can understand what you expect of them, and see how what they do contributes to its accomplishment.

When you look at the lists of the most admired, the most profitable, and the best firms to work for, you are often struck by the number of companies that appear on all three lists. This is not a coincidence. Everyone at those firms from the CEO on down knows what is important to their business's success and can see how what they do contributes to its accomplishment. The answers are normally not very long or complicated, but they are uniform in their content. This means precious time is not lost figuring out what to do and how it should be done. When employees are confident they know what this is, they quickly go about doing it without fear that some superior will contradict them.

Building the Organizational Structure

With the purpose and objectives firmly implanted, management needs to turn its attention to developing a system of interlocking jobs that will turn the plans into reality. The first step is defining all the individual tasks that need to done, and then grouping compatible tasks into jobs. A critical part of this process is to define the critical tasks. **Critical Tasks** are those things that must be done well for the firm to survive, or if done poorly will cause the firm to fail. A good example of a critical task in an agribusiness is proper food handling. Recently several fast food restaurants found that failure to do this properly can lead to the death of customers from E coli infections and bankruptcy of their firm. Critical tasks must always be given top priority in a business. They can never to subordinated to other tasks for the sake of efficiency.

The Elements of a Good Motivational Environment

With the foundation in place, managers can turn their attention to developing a good motivational environment where their workers will seek to excel. To do this manager must:

- Mesh the individual tasks into jobs and the jobs into a system that actually produces a product or service.
- Develop this system of jobs with customer, not employee needs, foremost in mind.
- Provide each employee with
  --proper tools, equipment, and training
  --clear, concise instructions on how to perform the job
  --a clear concise understanding how each job helps the firm to accomplish its objectives
- establish pay commensurate with the job and its performance
- establish job performance standards using measurable criteria such as cost/pound, pounds produced/week, or loads/day.
- give frequent and meaningful feedback to employees on their performance using the performance standards set above rather than on other, unnamed items.
- establish and maintain clear lines of authority between all levels of management.

This may seem like a lot to ask of any manager, but if you look at successful businesses they do this every day and in every way. This approach has many advantages. First, by leaving the details of how to accomplish their work to the employees, workers get to bring to bear their own special insights in how best to do their job. The results may exceed the manager's expectations. Second, this approach gives workers a feeling of control over their work and that what they say can have some impact. Third, this participatory approach can lead to greater profits.

Developing Your Motivational Skills

Besides the items above managers must also develop their motivational skills. These include:

- Understanding and putting a value on the part that each employee plays in the overall accomplishment of the firm's objectives
- Knowing how well the people you supervise are doing and giving them frequent feedback
- Communicating to workers that you know them, understand their concerns, and accept them as worthwhile people
- Establishing a feeling of trust and security between yourself and your employees. This means always being truthful and willing to defend them from unwarranted attacks by others
- Listening and actively soliciting the opinions of your employees
- Providing a challenging work environment
- Applying all the rules rewards, and penalties fairly to all

Summary

Building and maintaining a good motivational environment takes times. Unless everyone, especially top management, accepts this process and goes about it honestly, the firm's objectives will never be reached. Employees can quickly spot those just going through the motions. Successful development of a good motivational environment requires considerable effort and can be maintained only through consistent personal effort by all members of management every day.

References

Beierlein, James G., Kenneth C. Schneeberger, and Donald D. Osburn, *Principles of Agribusiness Management*, Waveland Press, 2nd Edition, 1995.

Leadership: Coaching to Develop People

Robert A. Milligan, Ph.D.
J. Thomas Clark Professor of Entrepreneurship and Personal Enterprise
Department of Agricultural, Resource, and Managerial Economics
Cornell University

—*Robert A. Milligan's speaker biography appears on page 133*—

Leadership by its name and common understanding means to lead. Today, however, leadership still means to lead but that is only a part of the job. Leading especially in the establishment of and commitment to the mission, vision and core values of the organization remains on the job description. New to the job description but perhaps equally as important is developing others, ideally everyone, to be a leader.

In their national bestseller <u>Flight of the Buffalo: Soaring to Excellence, Learning to let Employees Lead</u> James Belasko and Ralph Stayer explain why this change in leadership reflects the realities of today's world. They argue our traditional paradigm, based on capital as scarce resource, had the following characteristics.

- Markets were local or national.
- Work was unskilled and manual.
- Stability was the rule.
- Workers were uneducated.
- Communications took days or weeks.

Today, they argue the principal tools of production are ideas and talents. They call this intellectual capitalism; it has the following characteristics:
- Markets are global.
- Electronic highways enable instant communication and rapid competitive responses.
- Work involves the creation, transmission and manipulation of information and knowledge.
- Workers are highly educated.

In this environment the leaders role becomes one of provide the culture where every employee acts like they own the business and they developing their capabilities to the fullest. Belasco and Stayer use the analogy of the buffalo herd, which was completely lost with out its leader, as the old paradigm and a flock of geese, where the lead duck is rotated to capitalize on the strength of each duck, as the analogy for the new paradigm.

Belasco and Stayer suggest that the following four activities now constitute the job description for a leader:

1. Transfer ownership for work to those who execute the work.
2. Create the environment for ownership where each person wants to be responsible.
3. Coach the development of personal capabilities.
4. Learn fast and encourage others to learn quickly.

The outline below provides additional information on the first three components of the job description. The fourth is more self explanatory involving leaning and continuous improvement.

Transfer ownership for work to those who execute the work
- Switch mindset from problem solver to problem conveyer -- very difficult.
- Must help people learn "I am responsible" behavior meaning "I am able to respond effectively and appropriately."
- The person doing the job must own the responsibility for doing it correctly.

Create the environment for ownership where each person wants to be responsible
- Leader must create the environment where employee can own the responsibility for delivering great performance.
- "Owners" require information to make decisions.
- Must pay for ownership behavior.
- "If you want them to act like it's their business, make it their business."

Coach the Development of Personal Capabilities
- Coaching is about providing support and guidance.
- Develop individuals skills and competencies.
- Everything begins with delighting the customer – "From the customer's point of view, what is great performance?"
- Coaches raise expectations.
- Focus on developing the person, not the scorecard.

An important skill in leadership to develop people is to recognize different leadership styles and be able to select and utilize the appropriate style depending upon the situations, the individual you are leading and the relationship you have with that person. Six leadership styles constitute our leadership options. They are :

Coercive: Managers with this style tend to expect immediate compliance with their directions and solicit very little to no input. They manage by controlling subordinates tightly and tend to influence with discipline.

Authoritative: Managers who use this style are often referred to as "firm but fair." They tend to manage by providing clear instruction, soliciting some input (while leaving no doubt as to who is boss), monitoring behavior, and motivating with both discipline and rewards. They see influence as a key part of the manager's job.

Affiliative: Managers with this as their dominant leadership style tend to feel people come first and tasks second. They see the manager's job as one of maintaining a pleasant working environment and providing job security and other benefits. They want to be liked and they tend to provide little direction, especially feedback about unsatisfactory performance.

Democratic: These manager's are known for their participative style. They tend to believe that individuals and groups function best when allowed to work together and, therefore, tend to feel that close supervision or very detailed instructions are not necessary. Democratic managers tend to hold many meetings, they reward adequate performance, and they dislike disciplining employees.

Pacesetting: These managers like to perform technical activities as well as manage. They have very high standards for themselves and expect the same of others. These managers usually expect their employees to develop a keen sense of personal responsibility. They often have little concern for interpersonal relations and may reassign work if employee ability or willingness hampers performance.

Coaching: Managers using a coaching style see themselves as developing their employees and have high standards of performance. They delegate authority and allow followers flexibility in setting goals and completing tasks. They provide strong support when needed.

The table on the following page summarizes several advantages and disadvantages of each style.

References

Belasco, James A. and Ralph C. Stayer. Flight of the Buffalo: Soaring to Excellence, Learning to Let Employees Lead, 1993, Warner Books, Inc., New York

Maloney, Thomas R. and Robert A. Milligan. 1995. Human Resource Management for Golf Course Superintendents.

LEADERSHIP STYLES - ADVANTAGES AND DISADVANTAGES

Coercive

Advantages:
- Short term efficiency - fast
- Clear line of authority - know who is in charge and desired action is usually taken

Disadvantages:
- Most people don't like it
- Inhibits employee growth and development
- May lead to high staff turnover

Authoritative

Advantages:
- Efficient
- Clear who is in charge
- A way of exercising power without intimidation

Disadvantages:
- Not conducive to personal growth and development
- Some people may not like it
- Could lead to turnover

Affiliative

Advantages:
- Keeps people happy (short term)
- Allows people freedom

Disadvantages:
- Change is avoided and becomes a source of conflict
- Low productivity
- Decisions may not be in best interest of organization
- Little encouragement for personal growth

Democratic

Advantages:
- Involves people
- Opportunity for personal and organizational growth

Disadvantages:
- Time consuming
- Losers may sabotage organizational goals
- Majority decisions aren't always in best interest of the organization

Pace Setting

Advantages:
- Can be very productive (short term)
- Works well with committed followers

Disadvantages:
- Doesn't work well with unwilling or unable followers
- Followers may not follow
- Problem in absence of the pacesetter

Coaching

Advantages:
- Encourages growth and development
- Long-term productivity
- People respond well

Disadvantages:
- Time consuming
- Costs are high if you have turnover due to the development investment that is lost

Hiring with and without a Contract

Jennifer LaPorta Baker
Attorney
McNees, Wallace, and Nurick
Labor and Employment Law Practice Group
100 Pine Street/P.O. Box 1166
Harrisburg, Pennsylvania 17108
(717) 237-5209

— Jennifer LaPorta Baker's speaker biography appears on page 127 —

Most state courts, including Pennsylvania, have presumed that employees work "at will" and are retained "at will." What this means is that employees typically can quit for any reason or no reason, and that they generally may be discharged for any reason or no reason. The significance of this presumption cannot be missed. Unless the presumption is overcome, an employee typically has no legal basis to challenge his or her termination from employment. Nevertheless, to understand the power of the presumption, an agricultural employer also must recognize its "exceptions."

I. The Employment At-Will Doctrine

The doctrine of employment-at-will provides that employees may be discharged "with or without cause, at pleasure, unless restrained by some contract . . . ." Therefore, as a general rule, there is no common law cause of action against an agricultural employer for termination of an at-will employment relationship. The courts of most states have recognized exceptions to this rule only in the most limited of circumstances.

II. Exceptions to the At-Will Doctrine - Wrongful Discharge

Although an employer may discharge an at-will employee for good, bad, or no reason, it may not discharge for an unlawful reason. At-will employees have attempted to bring wrongful discharge claims based on the theories outlined below.

A. Statutory Claims of Discrimination

Where a specific statute prohibits an employer from basing a decision to discharge on a certain criteria, even an at-will employee can challenge a discharge based on the unlawful factor. Various state and federal anti-discrimination laws prohibit an employer from discharging an individual on the basis of certain factors, such as race, sex, religion, disability, etc. If an employer discharges any employee, including an at-will employee, on these grounds, the decision is unlawful and the employee typically may challenge the decision under the anti-discrimination laws that specifically permit a cause of action. Similarly, laws such as the Fair Labor Standards Act and the Family and Medical Leave Act specifically permit an employee, even an at-will employee, to bring a suit against the employer if the employer terminates the employee for exercising rights guaranteed under the particular statute.

Examples of other conduct protected by a statute includes the following:

- **Union Activity** -- The National Labor Relations Act prohibits the discharge of an employee because the employee engaged in union activity or other protected "concerted activity."

- **Wage Payment Complaints** -- Federal and state law prohibit an employer from terminating an employee for filing a complaint or testifying against an employer who is alleged to have violated wage and hour laws.

- **Safety Complaints** -- Federal and state law prohibit an employer from terminating an employee because he has filed a complaint with a federal or state agency complaining about unsafe working conditions.

- **Polygraphs** -- Federal and state law prohibit an employer from terminating an employee for refusing to take a lie detector test.

- **Jurors and Witnesses to Crimes** -- State law prohibits an employer from terminating an employee for attending jury duty or attending court by reason of being a victim of, or a witness to, a crime.

- **Criminal History** -- State law prohibits an employer from terminating an employee on the basis of a prior arrest record or a criminal conviction if the conviction is not relevant to suitability for the job.

Of course, a variety of civil rights acts also prohibit termination or any other form of employment discrimination which is based on an individual's membership in a "protected class." Currently, the major classes protected in Pennsylvania relate to a person's race, color, religious creed, ancestry, age, sex, national origin, non-job related handicap or disability, or the use of a guide or support animal because of handicap or disability. The administrative body that employers are likely to come into contact with when a claim of discriminatory termination is made is the Equal Employment Opportunity Commission (EEOC), although state and local statutes also proscribe discriminatory conduct by employers.

Various states have passed "whistleblower laws" prohibiting an employer from discharging an employee for reporting an act of wrongdoing by the employer or its agents. These laws constitute yet another restriction on the ability to terminate employment, including at-will employment. For example, the Pennsylvania General Assembly has promulgated a Whistleblower Law protecting employees who make a good-faith report of wrongdoing or waste to his or her superiors, to an agent of the employer, or to an appropriate governmental authority. This statute, however, has been interpreted to apply to public employees only.

B. Public Policy Claims

Even where a statute does not specifically make a particular reason for termination unlawful, courts have permitted at-will employees to bring a suit of wrongful discharge or wrongful termination claim against their employer where the employer's decision to terminate constituted a "violation of public policy." For example, under Pennsylvania law, a person is required to serve as a juror when called. Therefore, if an employee is discharged for properly responding to the call to serve as a juror, the employer's decision may be viewed as violating the clearly enunciated public policy favoring jury service and may be deemed unlawful. Other examples of situations where the courts have found violations of public policy include the following:

- Discharge for failure to take a polygraph test in violation of state statute.

- Discharge for filing a worker's compensation claim.

- Discharge for reporting or objection to employer's violation of federal laws and regulations.

- Discharge for filing an unemployment compensation claim.

It must be emphasized, however, that "public-policy" exceptions to the employment-at-will doctrine are extremely rare in many states, especially in Pennsylvania.

C. Termination with Specific Intent to Harm

Agricultural employers must be aware that the manner in which an employee termination is carried out may result in tort claims in certain situations. As a general rule, most states do not recognize a cause of action for wrongful discharge based on an employer's specific intent to harm an employee. Nonetheless, employers must be aware that their activities could result in tort claims being filed against them in specific situations. A classic example of such a situation is where, in a claim of sexual harassment, an employer disciplines the alleged harasser without conducting a thorough investigation and makes the reasons for the discipline known to persons other than the alleged victim. Employers should also be mindful of the fact that providing an unsubstantiated, negative employment reference regarding former employees may also give rise to a claim of defamation.

D. Civil Conspiracy to Wrongfully Terminate Plaintiff's Employment

Plaintiffs frequently include a civil conspiracy count in their wrongful discharge complaints, alleging that their supervisors conspired with the employer in the furtherance of their discharge. In Pennsylvania this claim is legally deficient because under the definition of civil conspiracy, a corporation cannot conspire with its agents/employees nor can agents of a single entity conspire among themselves. Further, since conspiracy requires an illegal act, there can be none where the employee is terminated for a lawful reason.

E. Misrepresentation

Statements made during the hiring phase of employment may constitute misrepresentation if they are a misrepresentation of a material fact, the defendant knew the statement to be false (or had a reckless disregard for the truth), the defendant intended the statement to induce the plaintiff to act, the plaintiff justifiably relied on the misrepresentation and the plaintiff was damaged by the misrepresentation. Example: if a recruiter lies about a substantial bonus the applicant will receive if hired in order to induce the applicant to leave his/her other job and the applicant does indeed leave his/her job in reliance on the promise of a bonus.

Negligent misrepresentation is a theory that has gained far less acceptance than fraudulent misrepresentation.

F. Intentional Interference with Contractual Relations

Another common count in wrongful discharge complaints is a claim for intentional interference with contractual relations. Just as in the conspiracy theory discussed above, the claim is flawed if the alleged interference was perpetrated by the plaintiff's managers and supervisors.

III. Contractual Challenges to Employment At-Will

If an employee's employment relationship is governed by a contract which may not be terminated except for specific reasons, such employment is no longer "at-will" but is governed by the terms of the contract. For this reason, many former employees attempt to establish that their employment was in fact governed by some form of contract which limits the employer's ability to terminate the relationship.

A. Express Contract Theory

The presumption of at-will employment can be overcome by the existence of an express contract for employment, oral or written. In order for a contract to be sufficient to overcome the presumption, however, it must be clear and definite and must contain a **specific** term of employment, such as a period of years. Pennsylvania courts typically have found that promises of lifetime or permanent employment are too indefinite to create a contractual relationship. Where an individual is retained pursuant to a valid contract of employment for a definite term, he or she may bring a breach of contract action against the employer for a discharge that violates the provisions of the contract.

B. Implied Contract Theory

Where an express contract does not exist, an employee nevertheless may attempt to overcome the at-will employment presumption by claiming the existence of an implied contract. An at-will employee who brings a claim under the implied contract theory is faced with a difficult proposition indeed. Such employee must essentially show that the employment relationship was not of an at-will nature at all. Rather, he must prove that a contractual relationship existed between himself and the employer even in the absence of a specific document to support his claim. The more popular of the implied contract theories are (1) implied contract based upon statements contained in employee handbooks and/or personnel policies; and (2) implied contract based upon additional consideration.

- Handbooks or Personnel Policies. Employees often will bring a breach of contract claim against the employer and assert that certain employee policies (e.g., progressive discipline schedules), or even an employment handbook, created a contractual relationship that legally is binding on the employer and the employee and not capable of breach.

- Additional Consideration. Many courts have been willing to imply a contractual employment relationship for a reasonable period of time where an employee alleges that additional consideration was given to his employer. The additional consideration may be in the form of a substantial benefit to the employer other than the services which the employee was hired to perform or a substantial hardship to the employee. For example, in a case where the plaintiff left his job as vice president of human resources at a hospital in Virginia for a similar position at a hospital in Pittsburgh, the jury found in favor of the plaintiff on his claim for breach of implied contract. The Pittsburgh hospital had induced the plaintiff to leave his job with promises of future opportunity. Sixteen days after he began his employment, the plaintiff was fired. The jury found that an implied contract existed and held that the hospital breached its contractual obligations. It found that additional consideration existed because the employee quit his secure, high paying job, sold his house and moved to Pittsburgh with his pregnant wife and two-year old son. Therefore, the jury found that the plaintiff relied to his detriment on the employer's offer of employment and promises of future opportunities. In another case, the plaintiff reluctantly left his previous employment, sold his house, and relocated; the court affirmed a jury finding of additional consideration after that individual was fired after only three months of employment.

C. Promissory Estoppel Theory

Although Pennsylvania does not recognize the doctrine of promissory estoppel, some states do. This doctrine simply stated is that an employer is prohibited from firing an employee for relying on the employer's alleged promise.

D. Good Faith and Fair Dealing

Although many plaintiffs continue to include this theory in their complaints, most states do not recognize an implied covenant of good faith and fair dealing in the employment relationship.

E. Collective Bargaining Agreements

Collective bargaining agreements between unions and employers typically contain language that provides that employees will not be disciplined or discharged except for "just cause." Covered employees, therefore, are not at-will employees since there exists a contractual restriction on the employer's ability to terminate the employment relationship. Typically, employees covered by a collective bargaining agreement with "just-cause" language are able to challenge a discharge decision through a contractual "grievance procedure," the final step of which ordinarily requires arbitration before a neutral third party. The third party would hear the facts of the discharge, the employer's position, and the employer's reason for the discharge and make a decision of whether "just cause" existed to support the decision. The employee, union, and employer typically agree to be bound by this decision.

IV. Maintaining an At-Will Relationship

Employers who wish to maintain an at-will employment relationship must at a minimum:

- Include an at-will disclaimer in employment applications and employee handbooks which states that the employment relationship is at-will and can be terminated at any time without prior notice or cause. It should be noted that any employee policies should state that the policy is only guidance and the employer reserves the right to amend or terminate such policy at any time.

- Avoid doing and saying things which either create an express or implied contract such as statements that employees will remain employed for a definite duration of time or that they will not be discharged without just cause. Such statements are often mistakenly made in written personnel policies and/or employee handbooks, to applicants during the recruiting stage or in the initial interview, to employees during performance evaluations, and in the regular course of employment by supervisors wishing to reward a job well done. As a general rule, employers should not promise anything regarding job security, tenure, or continued employment unless they are prepared to live up to their promises. Employers must specifically avoid statements such as the following:

    "you will become permanent after you pass probation";

    "you will always have a job here as long as you perform satisfactorily";

    "we give raises every year for good job performance";

    "if you come to work for us, you will be promoted to _____ within _____ amount of time"; and

    "you will never have to worry about a layoff here".

- Avoid unwarranted praise to employees, and if performance evaluations are used, they should accurately reflect an employee's performance.

- Ensure that your employees understand what is expected from them from the beginning of the employment relationship.

V. Conclusion

Although employment at-will remains the law in most states, a prudent agricultural employer will make an effort to determine that appropriate grounds exist for either a disciplinary or discharge reason. This will help to ensure that there is a certain degree of fairness in the manner in which that decision is communicated to the employee. Even if an employer has attempted to hire employees as carefully as possible, there nevertheless may arise situations where the employment relationship must be terminated. The employer must know the fundamental legal principals at least well enough to know if legal assistance may be necessary. Implementing the lessons learned in these discussions will go a long way toward assisting the employer in avoiding the legal pitfalls of terminating an employee regardless of whether that employee was hired at-will or pursuant to an employment contract. This is especially true in today's litigious environment where most any employee can bring a claim against an employer upon termination of employment.

Elements of an Employment Contract

John C. Becker, J.D.
Professor of Agricultural Economics and Director of Research
The Agricultural Law Research and Education Center
The Dickinson School of Law
The Pennsylvania State University

— John C. Becker's speaker biography appears on page 128 —

Introduction

The decision to employ someone under the terms of a contract is an exercise that requires a full and complete understanding of the relationship between the parties and the duties and obligations that arise from that relationship. An implied step in this process is to assure that both the employer and employee know and understand what the contract provisions mean to each of them and that important choices were made in selecting specific provisions for the first draft of the agreement.

Preliminary Considerations

Why should an employer devote time and resources to developing an employment contract for some or all of his employees? Most employers would answer that question by saying that a contract offers the advantage of providing additional employment security to those employees who seek that in their employment relationship. As the earlier discussion of "at will" employment points out, an employer who provides additional security to an employee may give up some control over the employment relationship, such as limiting the circumstances or situations in which the employer can terminate the relationship. The central point is that both employers and employees seek to gain certain advantages by having an employment contract and some negotiation may be needed to determine which advantages are gained and which are negotiated away in favor of gaining more desirable advantages. Therefore, it would be wise for employers to evaluate the advantages they want to obtain through an employment contract and then prioritize the list to identify the most desirable advantages to incorporate into the agreement. Employees should also follow a similar process.

A second consideration is timing the negotiation. When should contract development occur? Should it be before or after the decision is made to offer employment to a person? Should the process of defining the contract be made at the same time as each party evaluates the other and decides if there is enough potential for a mutually rewarding opportunity to merit pursuing negotiations further. To some employers, negotiations with prospective employees may be unfamiliar and, therefore, uncomfortable.

A third preliminary issue is that the law of employment contracts generally imposes requirements and conditions on enforcement of specific contract terms[1]. Simply having a provision in the contract does not assure it will be enforced by the courts as it is written. This is particularly true of provisions intended to restrict or confine an employee's ability to seek employment with another employer in the area or to open a new business in direct competitor with the initial employer. This consideration may lead an employer to decide that an employment contract may only be appropriate for some, but not all, employment situations, such as management or supervisory positions where the need to retain key employees may be greater than it would be for general workers or laborers.

Key Questions to Answer Before Drafting the Agreement

Is an employer-employee relationship the most advantageous relationship for the employer? Will another type of relationship arrangement, such as an independent contractor relationship be more effective[2] ?

Is the employee currently employed under contract to another employer? If so, how does that contract affect the employee's ability to enter into this contract? Are there any restrictions or

---

[1] For example, the right to seek an employee's covenant not to compete by working for a competitor, or opening a new competing business must meet four general conditions; (1) the restrictions must relate to the employment contract; (2) the conditions must be supported by consideration; (3) the restriction must protect a legitimate interest of the employer, and (4) the restriction must be reasonably limited in territory and duration. Of these four, the most often disputed term refers to the time and duration of the restriction. In general courts have said that these restriction must be no longer and the area no greater than that which is reasonable necessary for the protection of the employer's interest. See Records Center, Inc. v. Comprehensive Management, Inc.. 363 Pa. Super 79, 83-84' 525 A.2d 433, 435 (1987).

[2] See generally, John C. Becker and Robert G. Haas, The Status of Workers as Employees or Independent Contractors, Drake Journal of Agricultural Law, Vol. 1, No. 1, pages 51-72 (1996).

Is the employee currently employed under contract to another employer? If so, how does that contract affect the employee's ability to enter into this contract? Are there any restrictions or covenants in that agreement that could be interpreted as preventing the employee from entering into this agreement?

Will drafting an employment agreement that specifically details the employee's duties be too restrictive on the employer's ability to assign additional duties to the employee?

On what basis will the employee s compensation be calculated? Will it be an hourly or salaried basis? What other employee benefits will the employee be entitled to receive?

How will the employer's interests be affected if the employee opens a competing business or leaves this employment for that of a competitor?

Will the employee come in contact with the employer's important proprietary information which, if it is released, could put the employer at serious disadvantage? What is the nature of this proprietary information and is it a legitimate interest of the employer that is worthy of being protected?

Will the employee be able to develop items which are eligible for patent or copyright protection? If so, who should receive the benefits flowing from development of these intellectual properties?

Will the employee have access to the employer's funds or assets? If so, is a fidelity bond appropriate?

Are the parties willing to enter into an agreement to arbitrate or mediate their disagreements that may arise under the terms of this agreement? If so, does either party favor one form of alternative resolution rather than another? On what basis should arbitration proceed, under common law arbitration rules or under a statutory form of arbitration, if available under applicable state law.

Will other special employment laws apply to the employment relationship, such as the Migrant and Seasonal Agricultural Worker Protection Act, or a state Seasonal Farm Labor Law ?

Does the employee need to have special licenses or qualifications to perform the essential duties of the position, such as a commercial driver's licence or certification as a pesticide applicator or a nutrient management specialist?

How will the employment agreement's compensation and benefit package provisions affect the employer's tax situation?

Checklist of Specific Contract Provisions

1. Identification of the parties by name and address.

2. Description of the employer's business, the employee's professional skill level or position and employee's desire to enter into employment with the employer.

3. Describe the term of the agreement. Note: Employment is presumed to be at-will. In order to rebut this presumption, the employee has the burden of proving that the agreement is one for either a specific duration, that the employee will be discharged only for "just" cause, that the employee offered additional consideration to the employer, or that applicable and recognized public policy calls for such a result. <u>Luteran v. Loral Fairchild Corp.</u> 455 Pa. Super. 364, 370, 688 A.2d 211, 214 (1997).

4. Describe the duties of the employee, and consider the need to reassign the employee to other duties or to other places of work. How much time is the employee to spend in this employment?

5. Describe how the employee is to perform the duties of this position?

6. Describe how the employee will be compensated for his or her work? Does it include overtime pay, bonus compensation, commissions, incentive pay and on what basis is the incentive earned? When will a salary level be reviewed? Will compensation continue during employment interruptions, such as vacation, holidays, illness, disability, military service?

7. Describe the other employment benefits an employee is entitled to receive, such as insurance, use of vehicle, bonus arrangements, options to buy into the business? Will the employee be entitled to reimbursement for expenses incurred while performing duties of the employer?

8. If the employee creates something that can be protected under patent or copyright provisions, who will control the benefits that flow from the employee's efforts?

9. Is the employee subject to a restrictive covenant prohibiting direct and indirect competition with the employer or direct or indirect solicitation of the employer's customers or customer base? What is the term and the area of the restriction?

10. Is the employee restricted from using any trade secrets or proprietary information obtained in the course of employment with the employer?

11. Under what circumstances can the employment relationship be terminated? What process must be applied to complete the termination? What effect will termination have on the employee's entitlement to particular benefits of the employment?

12. Do the parties accept arbitration as the means of resolving disputes in the interpretation of the contract.

13. Include general contract clauses such as the manner of giving notice to the parties, merger of all prior written and oral discussions into the agreement, stipulation as to applicable law, provisions for attorney's fees, and the amounts due deceased employees.

14. Limits on the employee's ability to bind the employer to an obligation created by the employee.

15. Complete signatures of all parties witnessed by at least two individuals.

Guest Workers in Agriculture: The H-2A Temporary Agricultural Worker Program

Al French
Coordinator, Agricultural Labor Affairs
Office of the Chief Economist
U.S. Department of Agriculture
http://www.usda.gov/oce/oce/labor-affairs/affairs.htm

— Al French's speaker biography appears on page 131 —

The H-2A Temporary Agricultural Worker program permits agricultural employers who can demonstrate a labor shortage to import foreign workers on a temporary basis under terms and conditions that will have no adverse effect on U.S. workers similarly situated. The program is authorized by the Immigration and Nationality Act (INA) as amended by the Immigration Reform and Control Act (IRCA) of 1986. The following information provides a basic summary of the process of applying for an H-2A certification and the employers' obligations in doing so. The U.S. Department of Labor regulations at 20 CFR Part 655, Subpart B govern the application process.[1]

Who May Apply

An agricultural employer who needs workers to perform labor or services of a temporary or seasonal nature may apply. The employer may be an individual proprietorship, an association of agricultural producers, a partnership, or a corporation. An authorized agent also may apply on behalf of the employer.

What to Submit

- Application for Alien Employment Certification (Form ETA 750, Part A. Offer of Employment);
- Agricultural and Food Processing Clearance Order (Form ETA 790);
- Attachments as appropriate to supplement information requested on the above forms; and
- Statement of authorization of agent or association, if applicable.

---

[1] http://www.doleta.gov/regs/cfr/20cfr/toc_Part600-699/0655_toc.htm

When to Apply

Applications must be filed with the appropriate U.S. Department of Labor (DOL), Regional Administrator (RA), Employment and Training Administration (ETA) and local office of the State Employment Service at least sixty (60) calendar days before the first date on which workers are needed. In order to avoid delays in obtaining workers, first time applicants should submit their applications well in advance of the 60-day deadline in the event modifications to the applications are required by DOL. If the application is acceptable, the RA will make a certification determination twenty (20) calendar days before the date on which the workers are needed.

How to Apply

Applications may be filed in person, mailed certified return receipt requested, or delivered by guaranteed commercial delivery to the appropriate RA and local office of the State Employment Service. If a labor certification is granted, it is the employer's responsibility to arrange for the admittance of aliens into the U.S. by filing a visa petition with the Immigration and Naturalization Service.

Conditions to be Satisfied

- **Recruitment:** The employer is required to engage in independent positive recruitment of U.S. workers. This means an active effort, including newspaper and radio advertising as directed by DOL in areas of expected labor supply. Such recruitment must be at least equivalent to that conducted by non-H-2A agricultural employers to secure U.S. workers.

- **Wages:** The wage or rate of pay must be the same for U.S. workers and H-2A workers. The rate must also be at least as high as the applicable Adverse Effect Wage Rate[2] (AEWR) or the applicable prevailing wage rate, or the Federal or State minimum wage, whichever is higher. Prevailing rates are established by the employment service for specific crop and labor market areas.

    Federal Unemployment Tax Act and F.I.C.A. taxes are not payable for alien H-2A workers (because they are not eligible for benefits under these programs).

- **Housing:** The employer must provide free, inspected and approved housing to all workers who are not able to return to their residences the same day. Prospective H-2A employers should contact the Wage and Hour Division of the U.S. Department of Labor to arrange for an inspection at least 30 days prior to the date of their need for workers.

- **Meals:** The employer must provide either three meals a day to each worker or furnish free and convenient cooking and kitchen facilities for workers to prepare their own meals. If

---

[2] 1998 AEWRs for DE, MD, NJ, PA = $6.33; NY = $6.84; VA $6.16; WV = $5.92
http://www.usda.gov/agency/oce/oce/labor-affairs/aewr98.htm

meals are provided, then the employer may charge each worker a certain amount[3] per day for the three meals.

- **Transportation:** The employer is responsible for the following different types of transportation of workers: (1) After a worker has completed fifty percent of the work contract period, the employer must reimburse the worker for the cost of transportation and subsistence from the place of recruitment to the place of work. (2) The employer must provide free transportation between any required housing site and the worksite for any worker who is eligible for such housing. (3) Upon completion of the work contract, the employer must pay for return transportation or transportation to the next job.
  If it is the prevailing practice for employers in the same crop and labor market area to advance transportation to prospective workers, H-2A employers must also arrange or advance such transportation.

- **Workers' Compensation Insurance:** The employer must provide Workers' Compensation or equivalent insurance for all workers. Proof of insurance coverage must be provided to the RA before certification is granted.

- **Tools and Supplies:** The employer must furnish at no cost to the worker all tools and supplies necessary to carry out the work, unless it is common practice for the worker to provide certain items.

- **Three-fourths Guarantee:** The employer must guarantee to offer each worker employment for at least three-fourths of the workdays in the work contract period and any extensions.

- **Fifty Percent Rule:** The employer must employ any qualified U.S. worker who applies for a job until fifty percent (50%) of the contract period has elapsed.

- **Labor Dispute:** The employer must assure that the job opportunity for which the employer is requesting H-2A certification is not vacant due to a strike or lockout.

- **Certification Fee:** A fee will be charged to an employer granted temporary alien agricultural labor certification. The fee is $100, plus $10 for each job opportunity certified, up to a maximum fee of $1,000 for each certification granted. There is an additional petition fee payable to the Immigration and Naturalization Service.

- **Other Conditions:** The employer must keep accurate records with respect to a worker's earnings. The worker must be provided with a complete statement of hours worked and related earnings on each payday. The employer must pay the worker at least twice monthly or more frequently if it is the prevailing practice. A copy of the work contract must be provided by the employer to each worker.

---

[3] $7.60 per day in 1998

Review of Application

Under normal circumstances, the RA will notify the employer in writing within seven (7) calendar days after receipt of an application, if the application is acceptable or needs modifications.

Recruitment of U.S. Workers

Upon receipt of an employer's application for temporary alien agricultural labor certification, the local State Employment Service office must promptly prepare a local job order and begin recruiting U.S. workers in the area of intended employment.

Within seven (7) calendar days after receipt of an application, the local office must prepare an agricultural clearance order to permit the recruitment of U.S. workers by the State Employment Service system on an intrastate and interstate basis.

After an application is accepted for consideration, the RA will provide direction to both the employer and the State Employment Service on specific recruitment efforts to be conducted. These efforts may require the employer to advertise and recruit in States regarded as labor supply States and the Commonwealth of Puerto Rico.

Notices of Acceptance

- Will inform the employer and the State Employment Service of specific efforts expected regarding recruitment of U S workers;

- Will require that the job order be placed into appropriate intrastate and interstate clearances; and

- May require the employer to engage in independent recruitment efforts within a multi-state region (including the Commonwealth of Puerto Rico) of traditional or expected labor supply.

Notices that Applications are Not Accepted for Consideration

- Will state why the employer's application is not acceptable;

- Will state changes necessary for the application to be accepted for consideration;

- Will allow the employer five (5) calendar days to resubmit the application;

- Will outline procedures employer may use to appeal the RA's nonacceptance.

Basis for Denying Certification

- Insufficient time to test the availability of U S. workers;

- U.S. workers are available to fill all the employer's job opportunities;

- Employer has not complied with the workers' compensation requirements;

- Employer has not complied with recruitment requirements;

- Employer, since the application was accepted for consideration, has adversely affected the wages, working conditions, or benefits of U S. workers; and

- RA determines that the employer has substantially violated a material term or condition of a previous certification within the last two years.

Appeals of Denials of Certification

If it is determined that the employer has complied with the H-2A requirements, the RA will grant the temporary alien agricultural labor certification for the number of job opportunities for which it has been determined that there are not sufficient U.S. workers available. After certification, the employer must continue to recruit U.S. workers until the H-2A workers have departed for the place of work. In addition, the State Employment Service will continue to refer to the employer U.S. workers, who apply during the first fifty percent of the contract period, and the employer must hire these U.S. workers.

Violations, Penalties and Sanctions

A major consideration of IRCA is the enforcement of all provisions related to protections for workers. The Employment Standards Administration (ESA) of DOL has a primary role in investigating the terms and conditions of employment. ESA is responsible for enforcing contractual obligations of employers, and may assess civil monetary penalties and recover unpaid wages. ETA will enforce other aspects of the laws and regulations and will be responsible for administering sanctions for violations of the regulations.

Appeals of Employer Penalties

The RA will inform the employer about the system of appeals provided for in the regulations.

Where to Apply

Applications must be filed with the U.S. Department of Labor Regional Administrator, Employment and Training Administration in the region of intended employment. The Regional Offices are:

Regional Administrator, U.S. Department of Labor, ETA
Room 1707 J. F. Kennedy Bldg.
Government Center
Boston, MA 02203
(617) 565-2258

Regional Administrator, U.S. Department of Labor, ETA
201 Varick St.
New York, NY 10014
(212) 337-2139

Regional Administrator, U.S. Department of Labor, ETA
P. O. Box 8796
Philadelphia, PA 19101
(215) 596-6336

Regional Administrator, U.S. Department of Labor, ETA
Room 400
1371 Peachtree St., NE
Atlanta GA 30367
(404) 347-4411

Regional Administrator, U.S. Department of Labor, ETA
230 So. Dearborn St., 6th Floor
Chicago, IL 60604
(312) 353-0313

Regional Administrator, U.S. Department of Labor, ETA
525 Griffin Street
Dallas, TX 75202
(214) 767-8263

Regional Administrator, U.S. Department of Labor, ETA
Room 800 Federal Building
911 Walnut Street
Kansas City, MO 64106
(816) 374-3796

Regional Administrator, U.S. Department of Labor, ETA
1961 Stout Street
Room 1676
Denver, CO 80294
(303) 844-4477

Regional Administrator, U.S. Department of Labor, ETA
71 Stevenson Street
8th Floor
P. O. Box 3767
San Francisco, CA 94105
(415) 995-5482

Regional Administrator, U.S. Department of Labor, ETA
1145 Federal Office Building
909 First Avenue
Seattle, WA 98174
(206) 442-7700

Discrimination in the Workplace

Michael D. Pipa
Esquire
Mette, Evans, and Woodside
Harrisburg, Pennsylvania

— Michael D. Pipa's speaker biography appears on page 135 —

INTRODUCTION

Almost every employment relationship is subject, in some way, to prohibitions against discrimination under Federal and State employment laws. Under these laws, it is essentially illegal for employers to base employment decisions upon differences in race, sex, religion, national origin, age, or medical condition. This part of the seminar is intended as only a brief introduction into some of the major federal legislation. You should not underestimate the extent to which dissatisfied employees may claim discrimination. Remember also that this area of the law is developing rapidly, so that new laws are being enacted and existing laws amended frequently. In addition, most states have discrimination laws that closely follow the federal laws and that apply to small businesses.

ANTI-DISCRIMINATION LAWS

I. **Title VII**

    A. **Introduction**:

The full name of this law is "Title VII of the Civil Rights Act of 1964." The law is known generally as "Title VII." This law is the basis for most anti-discrimination laws in the United States. It has been amended by the Equal Employment Opportunity Act of 1972, the Pregnancy Discrimination Act of 1978, and the Civil Rights Act of 1991. The law essentially prohibits discrimination in employment based upon race, color, sex, national origin, and religion.

B. **Who Must Comply:**

Title VII has broad applicability and imposes limitations upon federal, state and local governments as well as private employers, labor organizations, and employment agencies. Covered employers include corporations, partnerships, and any other legal entity whose business affects commerce, and who has fifteen or more employees for each working day in the preceding twenty or more calendar weeks of either the current year or the preceding calendar year. There are limited exclusions for certain groups, such as bonafide membership clubs, Indian tribes and religious organizations. Title VII may even apply outside the territorial jurisdiction of the United States.

C. **Protected Employees:**

Title VII prohibits employers from making distinctions among employees on the basis of race; color; sex; national origin; and religion. The prohibitions against discrimination based on race or color extend to all races and colors, including Latinos, Asians, Native Americans, and Caucasians. The prohibition against religious discrimination includes not only established and organized faiths, but also those with sincerely held beliefs that have the strength of traditional religious views.

D. **Illegal Employment Actions:**

Title VII contains the very broad statement that it is unlawful for an employer to "fail or refuse to hire or discharge any individual, or otherwise to discriminate against any individual with respect to his compensation, terms, conditions, or privileges of employment." Under this law, therefore, virtually every decision made and action taken by an employer is subject to scrutiny. Hiring and firing decisions are the obvious areas of concern, but the anti-discrimination laws extend to all terms and conditions of employment, including disciplinary actions and employee benefits packages.

Generally, complaints arise when an employer **discriminates directly** by, for example, refusing to hire women or minorities for certain jobs or providing lesser fringe benefits to females or minority employees. In order to recover, the employee must prove that race, color, religion, sex, or national origin was **a motivating factor** in the employment decision, not the only factor.

In other cases, an employer may be found liable even without evidence of direct discrimination if the members of a group can establish a **pattern or practice** of adverse or unequal advancement by certain groups, even though there is no outright, direct policy. Such cases are based upon historical and statistical evidence of discrimination.

1. **Sexual Harassment:**

Under Title VII, cases involving allegations of sexual harassment tend to be the most common. Allegations of sexual harassment may arise even where there is no direct or indirect policy of refusing to hire, or of different terms and conditions for members of different sexes. Specifically, certain conduct toward members of the opposite sex, even conduct that might have been acceptable in the past or with others, can be found to be discriminatory and in violation of the law. The courts have classified sexual harassment cases into two categories:

   a. "Quid Pro Quo": The Equal Employment Opportunity Commission (EEOC) guidelines on discrimination because of sex state generally that "unwelcome" sexual advances and requests for sexual favors, and "other verbal or physical conduct of a sexual nature" are considered sexual harassment when (1) either explicitly or **implicitly**, submission to this sexual conduct is made a term or condition of a person's employment, or (2) rejection of the sexual conduct results in adverse employment action. In other words, when a supervisor or manager grants job benefits in return for sexual favors, or punishes a subordinate for refusing to comply, Title VII is violated.

   b. "Hostile Environment": Sexual harassment can occur in cases where there is no direct benefit or punishment for either engaging in or rejecting sexual conduct. Thus, Title VII violations may occur where there is a pattern of verbal or physical conduct of a sexual nature, and that conduct has either the purpose or **effect of** interfering with a person's work performance or creates an intimidating, hostile or offensive working environment. An employer can be liable for all types of harassment, even by co-workers, if the employer knew or should have known of the conduct and failed to take prompt, effective remedial action. A isolated incident usually does not give rise to liability - the harassment must be severe or pervasive enough to alter the conditions of employment. However, the test of whether an environment is hostile is based upon the objective view of the employee, not the employer's subjective opinion. Whether an environment is "hostile" or "abusive" is judged by all of the circumstances, including the frequency and severity of the conduct and whether it unreasonably interferes with work performance.

II. Age Discrimination:

A. Introduction:

The Age Discrimination in Employment Act of 1967, known informally as the "ADEA," is modeled after Title VII. The law applies to governments and private employers who employ more than fifteen employees. Employers are subject to liability both when they discriminate directly against older workers and also when, historically and statistically, their policies favor younger employees. Like Title VII, the ADEA prohibits discrimination in all aspects of employment, including hiring, promotion, compensation, discharge, and working environment.

B. Protected Employees:

Persons who are at least 40 years old are protected under the ADEA. The ADEA does not prohibit age distinctions applied to persons under the age of 40.

C. Illegal Employment Actions:

You may not base employment decisions on old age. Fixed retirement ages, for example, will in most cases violate the ADEA. (There are, however, certain statutory exceptions and defenses under which an employer may be entitled to impose a fixed retirement age.) It is also unlawful under the ADEA to impose certain requirements or conditions upon older workers that are not imposed upon members of other age groups. For example, an employer who requires employees over a certain age, such as the age of 55, to take physical or mental tests, but does not require the same tests of employees under that age, violates the ADEA. Any decision that is motivated by age gives rise to liability. An employee must prove that age played a role, but not the only role, in an employment decision.

The ADEA does not prohibit decisions that the employer can show are motivated by factors that may correlate with age, such as years of service or pension status. However, employees may suspect that age is the real reason and may use such programs as circumstantial evidence of intentional age discrimination.

D. Defenses and Exceptions:

The ADEA was enacted to protect older workers against intentional discrimination and policies based solely on the fact of old age. Employers may still have bona fide seniority systems, which in practical effect treat older workers differently, without violating the Act. An employer may also have job requirements based upon age if the employer can establish that age is a "bona fide occupational qualification." In order to prove this bona fide qualification, an employer must show a factual basis for believing that either all or substantially all persons over the age in question would be unable to perform essential job duties safely and efficiently. Decisions may also be based upon reasonable factors other than age, such as educational credentials and prior experience. However, these requirements must be applied uniformly to all employees, regardless of age. Employers may also employ and observe terms of certain bona fide benefit plans, such as retirement, pension, life insurance, etc., which in effect make age-based distinctions. However, in order to defend any such program, the employer must show that "for each benefit or benefit package, the actual amount of payment or cost incurred on behalf of an older worker is no less than that made or incurred on behalf of a younger worker."

III. Americans with Disabilities:

A. Introduction:

This is the law aimed at protecting employees who may have physical handicaps or disabilities. The full name of the Act is the Americans With Disabilities Act of 1990, and the law is known informally as the "ADA." The ADA also applies to employers with more than fifteen employees.

B. Protected Employees:

The ADA was enacted in order to ensure that handicapped or disabled persons would have equal opportunities in the work place. Therefore, employers may not refuse to hire or otherwise discriminate against persons with disabilities simply based upon the fact that the person is disabled. Since the ADA was enacted, one of the problems has been determining whether an employee or prospective employee is "disabled" and therefore protected by the ADA.

The ADA applies to those (1) with a physical or mental impairment that substantially limits one or more of the major life activities of an individual; or (2) who have a record of such impairment; or (3) are being regarded as having such impairment.

Therefore, an employer may not discriminate against a person who is limited currently from performing one or more of the major life activities; or who in the past had such an impairment; or who the employer "regards as" impaired. The "regarding as" category means that employees who are not truly disabled may sue, claiming discrimination by an employer who thought they were disabled.

The definition of disability is broad. Persons covered include those with disorders and losses affecting almost any body system, including mental or psychological disorders and emotional illnesses. Those with cardiovascular disease, epilepsy, dyslexia, Aids, manic depressive symptoms, and even unusual sensitivity to things like tobacco smoke may fall within the broad definition of a person with a "disability." On the other hand, "normal" physical characteristics, such as height and weight, and personality traits such as poor judgment or a quick temper are not disabilities.

C. **Illegal Employment Actions:**

Employers should not treat those with disabilities any different in regard to any term or condition of employment. Do not classify a disabled person in a way that would adversely affect job opportunities or status. An employer may also not use standards, tests, or other employment practices that, **in effect**, discriminate on the basis of disability. Similarly, an employer may not discriminate against a person because of that person's known relationship to or association with someone known to have a disability. Employers may not ask whether a job applicant has a disability or ask about an applicant's worker's compensation history. An employer may also not require pre-employment medical and psychological examinations unless they are designed to detect illegal drug use. Employers also may not discriminate because of the effect a disability may have on the employer's health and life insurance and other benefits. Employers may require medical examinations after an offer of employment has been made or after employment commences. However, such medical examinations must be required of all employees in that job category, the results must be kept confidential, and any criteria must be job-related and consistent with business necessity.

D. **Reasonable Accommodations:**

Not all persons with disabilities are able to perform in a job. If the disability affects an employee's ability to perform, the employer may take adverse employment action. However, the employer is first obligated to attempt a "reasonable accommodation" for any qualified individual with a disability. Generally, a reasonable accommodation includes modifications and adjustments to the job application and testing process (providing qualified readers or interpreters), modifications or adjustments to the work environment (such as buying a new chair, restructuring a job or work schedule, etc.) or assigning the employee to a vacant position.

However, an employer is not required to accommodate someone with a disability if the accommodation would impose an "**undue hardship**" on the operation of the employer's business. An undue hardship is an action that generally requires significant difficulty or expense - this is a fluid test that applies on a case-by-case basis. Relevant factors include the nature and cost of the accommodation, the financial resources of the facility, the number of persons employed, the effect of the proposed accommodation on the business resources, the size of the business in relation to the number of employees, and the type of operations covered including the composition, structure and functions of the work force.

In addition, an employer that does not know of a disability or the need for accommodation usually cannot be held liable. The covered individual must request an accommodation. All discussions with the employee should be based on job-related problems and issues, and not on the disability. If the employee refuses the accommodation proposed by the employer and otherwise cannot do the job, adverse employment action may be taken.

IV. **Equal Pay Act:**

    A. **Introduction:**

The Equal Pay Act of 1963 is an amendment to the Fair Labor Standards Act. The Equal Pay Act was enacted in order to ensure that male and female employees would be paid equally for "equal work."

    B. **Who is Covered:**

"Equal work" is not a term that can be defined precisely. Generally, the requirements of the Equal Pay Act apply to men and women who are in jobs that require equal skill, effort, and responsibility, and which are performed under similar working conditions. The Act does not apply to anything other than sex discrimination.

    C. **Illegal Pay Policies:**

Generally, the Equal Pay Act makes it illegal to pay male and female employees different amounts for equal work. The prohibitions also apply to fringe benefits. It is not illegal, however, for an employer to pay different wages based upon factors other than sex. Therefore, an employer may still use seniority, merit, or the quality or quantity of work as a basis for paying different wages. However, you should be careful to document these systems so that they do not appear to be mere excuses for unequal pay.

V. **State Anti-Discrimination Laws:**

Most states and even some larger cities have their own anti-discrimination laws to prohibit employment discrimination. For the most part, these laws are based upon the Federal laws discussed above. However, these state laws **often apply to employers with fewer than the minimum of fifteen employees** required for federal coverage. Many of the states have their own employment agencies, and the majority of charges received by the Federal EEOC are deferred to such state or local agencies for initial processing. Thus, anti-discrimination suits against employers often result in charges under both the federal and state laws. Many of the state agencies also have the power to issue "cease and desist" orders.

In Pennsylvania, the Pennsylvania Human Relations Act applies to Pennsylvania employers who employ four or more persons within the Commonwealth. The Act is enforced by the Pennsylvania Human Relations Commission.

You should be familiar with or consult your attorney about the possible anti-discrimination provisions of the laws in effect in your state and locality.

Farm Employment Rules and Regulations: What You Need to Know

Al French
Coordinator, Agricultural Labor Affairs
Office of the Chief Economist
U.S. Department of Agriculture
http://www.usda.gov/oce/oce/labor-affairs/affairs.htm

— Al French's speaker biography appears on page 131 —

HIRING PROCEDURES CHECKLIST

Employee obligations:

- Complete employer's standard application form.
- Provide full name as used for social security purposes.
- Provide present address and permanent address.
- Provide Social Security number.
- Provide date of birth.
- Sex.
- Complete Internal Revenue Service Form W-4 "Employees Withholding Allowance Certificate."
- Complete applicant portion of INS I-9 employment eligibility verification form and provide sufficient documentation of employment eligibility.

Employer obligations:

- Provide to migrant workers, and on request to seasonal workers, MSPA terms and conditions of employment; optional form WH-516.
- If housing provided, have Federal and State inspection prior to occupancy.
- If housing is provided, provide terms and conditions of occupancy (Form WH-521).
- Obtain Social Security number.
    See "Social Security Number Verification" below.

- Forward copy of Form W-4 to your State Directory of New Hires (Personal Responsibility and Work Opportunity Reconciliation Act of 1996).
- If applicant did not present a EPA Worker Protection Standard Training Verification Card, arrange to provide to the new employee pesticide safety training.
- Check age of applicant against State and Federal child labor laws.[1]
- Do not treat applicants disparately on the application form or hiring process with respect to race, color, religion, gender, citizenship, national origin, age, or disability in order to comply with:
    Age Discrimination in Employment Act of 1967
    Americans With Disabilities Act of 1990.
    Civil Rights Act of 1964
    Equal Pay Act of 1963.
    Immigration Reform and Control Act of 1986.

- After commitment to hire, require completion of INS Form I-9.[2]
    Employer must accept documents listed on I-9 form if they reasonably appear to be genuine. These include:
        driver's license and social security card, or,
        for legal aliens, same as above or unexpired INS document.
    If employer believes applicant has invalid documents, provide an opportunity to correct the deficiency or produce other documents before refusing to hire applicant or terminating employee.
    Employer must not:
        Request specific documents.
        Request additional documents if those proffered are sufficient.
            However, employer may ask for social security card for tax filing purposes only, but only after I-9 process has been completed.

- If requested, employer must provide to workers Internal Revenue Service Earned Income Credit/Advance Payment Certificate (IRS Form W-5). If employee selects advance earned income tax credit and files the Form W-5, employer must compute and pay advance earned income tax each pay period and file the appropriate forms with the Internal Revenue Service for reimbursement.[3]

- Employer must display official posters as required by regulations listed below.

---

[1] http://hammock.ifas.ufl.edu/txt/fairs/19557

[2] http://www.usda.gov/agency/oce/oce/labor-affairs/ircadisc.htm

[3] See IRS Circular E, Publication 15, Employers Tax Guide and also IRS publication Earned Income Credit, Publication 596.

FEDERAL LAWS AND REGULATIONS

Note: There may exist State laws with requirements that exceed similar Federal laws.

Age Discrimination in Employment Act of 1967

The Age Discrimination in Employment Act of 1967 is intended to encourage the employment of persons 40 years of age and older based on ability rather than age.

Coverage: employers with 20 or more employees in at least 20 calendar weeks.

Americans With Disabilities Act of 1990.[4]

Coverage: employers with 15 or more employees.

The American s With Disabilities Act prevents discrimination against persons with disabilities in recruitment, pay, hiring, firing, promotion, job assignments, training, leave, layoffs, benefits, and all other job related activities.

Civil Rights Act of 1964

Coverage: 15 or more employees in 20 or more weeks of the current or previous calendar year.

Title VII of the Civil rights Act of 1964 prohibits employers from discriminating against employees based on race, color, religion, sex, or national origin. Employers may never discriminate on the basis of race or color, but may discriminate on the basis of religion, sex, or national origin if it is a bona fide occupational qualification. The employer has the burden of proving that to show that the job requirement is essential for the normal operation of the business. Such a decision should not be made without advice of legal counsel.

Equal Pay Act of 1963

Covers employers with 500 man-days of labor in any calendar quarter.

Different pay scales may be used only if made according to a seniority system, merit system, a system that measures earnings by quantity or quality of production, or a system based on any factor other than sex.

---

[4] http://janweb.icdi.wvu.edu/kinder/

Fair Labor Standards Act (FLSA) - the Minimum Wage Law[5]

The Fair Labor Standards Act (FLSA) provides covered employees minimum hourly wage standards which, beginning September 1, 1997, are $5.15 per hour. In addition, minimum age standards are provided.

Covered employers are those who used 500 man-days of labor in any calendar quarter of the previous calendar year. A man-day is any day in which at least one hour of work was performed

Overtime exemptions:

    Employees employed in agricultural activities such as cultivating the soil; growing or harvesting of crops; or the raising of livestock, bees, fur bearing animals, or poultry are exempt from the overtime provisions of the FLSA. Employees such as packinghouse or processing workers, or who work in the office, shipping, warehousing, transportation, and sales are exempt from overtime premium if they work solely with products grown by their employer. However, in any week such employees are engaged in handling product not grown by their own employer, time and one half for overtime hours worked is payable. Agricultural employers who may handle crops or nursery products produced by another farmer should seek legal advice or advice from the DOL Wage and Hour Division concerning specifics of this overtime provision.

    Employees engaged in the transportation of preparation for transportation of fruits and vegetables to the first place of processing or first marketing within the same State, even when performed for a non-farmer, are not eligible for overtime premium.

    The overtime exemption applies to an employee transporting harvest workers between any place and the farm in the same State even if the driver is not employed by the farmer.

---

[5] "Agricultural Employment Under the Fair Labor Standards Act," W.H. Pub. 1288. "Exemptions Applicable to Agriculture, Processing of Agricultural Commodities, and Related Subjects, Under the FLSA of 1938," W.H. Pub. 1042.

"Wage Payments Under Fair Labor Standards Act of 1938," W.H. Pub. 1210.

"Records to be Kept by Employers Under the Fair Labor Standards Act of 1938," as amended, W.H. Pub. 1261.

"Hours Worked Under the Fair Labor Standards Act, W.H. Pub. 1344, March 1976.

http://www.dol.gov/dol/esa/public/minwage/main.htm

Payroll Deductions.

It is permissible to make certain deductions which reduce wage level below the minimum wage. These include:

Taxes required by law, *e.g.*, F.I.C.A., withholding, Medicare.

Salary advances, but retain receipts.

Third party deductions authorized by the employee such as insurance premiums, merchant accounts, etc., so long as employer receives no benefit from the transaction.

Housing, utilities, and meals, provided the deduction does not exceed the lesser of actual cost or fair value and meets a number of specified conditions dealing with profit in 29 C.F.R., Part 531.

DOL has taken the position that if employers charge workers for housing that could be used only for migrant workers it does not have any fair rental value. Thus, charges for such housing may not lawfully reduce the pay below the minimum wage.

Employers may not make deductions that reduce pay below the minimum wage for:

Transportation advances. It is presumed that, if an employer makes such advances, it is for the employer's own benefit in order to assure a sufficient number of workers.

Charges for a FLC's services.

Charges for "tools of the trade and other materials incidental to carrying on the employer's business."

Employers must maintain records that include:

Full name of employee as used for social security purposes.
Permanent address of employee including zip code.
Sex and occupation in which employed.
The number of man-days worked each week or month (a man-day is any day in which one hour's work is performed).
Time of day and day of week on which workweek begins.
Basis (hourly, per day, piece rate) on which wages are paid.
Hours worked each day and total hours worked each pay period.
Total daily or pay period earnings

Total additions or deductions from wages with an explanation of each.
Total wages paid each pay period together with proof of payment to individuals including cash advances or other deductions.
Date of payment and pay period covered by payment.
Name, present address, permanent address and date of birth of any minor under 18 who works when school is in session.
(See also MSPA requirements for earnings statements for employees).

Employers must display official poster Notice to Employees which contains basic information on minimum wages.

Most Common FLSA Violations:

Failure to maintain accurate records.
Failure to post required notice.
Failure to properly pay employees for meal periods and waiting time.
   Work permitted during lunch period is hours worked..
Failure to pay employees for compensable training time.
Failure to pay employees for unauthorized but compensable overtime.
Child labor.
   If child is allowed to work, the child is an employee.
Deductions, for other than taxes or payday advances, which cause pay to fall below minimum wage. (Some narrow exceptions exist for other deductions).

Family and Medical Leave Act of 1993[6]

Coverage: Employers with 50 or more employees for 20 weeks in the year. Count employees of any farm labor contractor in determining coverage.

Requires employers to permit eligible employees to take up to 12 weeks of unpaid leave during a 12 month period for birth of a child, placement of a child for adoption or foster care; to care for a spouse, parent or child with a serious medical condition; and for employees own serious health condition that makes worker unable to perform job.

Employers must display poster; protect employee's job while on leave; maintain group health insurance coverage, if any; but may require employee to use available vacation or accrued sick leave before using FMLA leave.

To be eligible, the employee must have worked for the employer for at least one day in each of 12 months, but not necessarily consecutive months, and, must have worked at least 1,250 hours during the 12 months period preceding leave and, work within 75 miles of a worksite where at least 50 workers are employed by the employer.

---

[6]   http://www.usda.gov/agency/oce/oce/labor-affairs/fmla.htm

Field Sanitation (Occupational Safety and Health Act)[7]

Coverage: Employers who have employed 11 or more field workers at any time.

Exception for toilets and handwashing facilities if workers work less than 3 hours in a day (including transportation time); logging operations; or caring for livestock.

Provide toilets and handwashing facilities.
One for every 20 workers or fraction thereof.
Toilet facilities must be screened, insure privacy, maintained in sanitary condition, and have self-closing doors that may be latched.
Locate within one quarter mile of each worker.
Disposal of waste shall not create unsanitary conditions.

Provide potable drinking water which is readily accessible.

Sufficient, cool, clean, dispensed in single-use cups, and refreshed daily.

It is employer's responsibility to inform employees of the importance of drinking water frequently, handwashing before and after using toilet and before eating or smoking, and to urinate as frequently as necessary.

Housing

Coverage: Migrant labor housing built since April 3, 1980 is regulated by Occupational Safety and Health Administration regulations.[8] Housing built before April 3, 1980 and without major modifications is regulated by the less restrictive standards of the Department of Labor Employment and Training Administration.[9]

Housing may be inspected by DOL's Employment and Training Administration through State employment service agencies, DOL's Wage and Hour Division, and the Occupational Safety and Health Administration.

For purposes of housing regulations, the 500 man-day exemption of MSPA does not apply to employers who provide housing to migrant workers.

---

[7] C.F.R. 1910.128
http://hammock.ifas.ufl.edu/txt/fairs/19477

[8] Occupational Safety and Health Standards, 29 C.F.R. 1910.142
http://hammock.ifas.ufl.edu/txt/fairs/19520

[9] Part 654, Subpart E, Housing for Agricultural Workers, Employment and Training Administration, Federal Register, March 4, 1980, pages 1480-1486

Terms and conditions of occupancy (Form WH-521) must be posted or given to workers at the time of recruitment. State certificate of occupancy must be posted.

If housing is provided, have Federal and State inspection prior to occupancy.

Employers are not liable for lack of prior Federal inspection if the request for inspection was made at least 45 days in advance of occupancy.

Employers using the employment service interstate clearance system must have housing inspected and approved before completion of their application.

Employers are responsible for checking housing frequently to be sure it has not fallen out of compliance.

Housing providers must also display copy of State "Certificate of Occupancy."

Immigration Reform and Control[10]

Coverage: all agricultural employers and all agricultural workers.

It is unlawful for an employer to hire, recruit, refer for a fee, or continue to employ a person who is known (or should be known) to be an unauthorized alien.

It is unlawful to discriminate against any individual based on national origin or citizenship status.

Employers may rely upon a State employment service to complete I-9 forms for persons referred by the agency if the agency provides that service. Such certification must be retained the same as I-9 forms. A simple referral, without I-9 verification, from the employment service does not relieve the employer from liability.

Employers must:

Require each applicant to complete INS I-9 employment eligibility verification form after commitment to hire.

Accept documents if they reasonably appear to be genuine. These include:

driver's license and social security card, or,
for legal aliens, same as above or unexpired INS document, or, other documents listed on I-9 form.

Inspect applicant's documents noting the ID number and expiration date.

---

[10] http://www.usda.gov/agency/oce/oce/labor-affairs/ircasumm.htm

An employer who inadvertently makes a technical or procedural error in the I-9 process is entitled to an opportunity to correct the error without penalty.

If it becomes known that an applicant has invalid documents, employer should provide an opportunity for employee to correct the deficiency or produce other documents before termination.

Employers are not required to complete Form I-9 for employees of independent contractors.

Employer must retain Form I-9 for at least three years and present them for inspection upon request. INS inspection officers are required to give at least three days advance notice before an employer inspection.

Employer must not:

> Request additional documents if those proffered are sufficient. However, employer may ask for social security card for the purpose of completion of tax documents only, but only after I-9 process has been completed.
>
> Engage in disparate treatment of job applicants or employees based on national origin or citizenship status..

Income Tax Withholding

> Exemptions:
>
>> Employee earns less than $150 in calendar year.
>> Employer's annual payroll less than $2,500.
>> Hand harvest employee if
>>> Paid on piece rate basis **and**
>>> Commutes daily from permanent residence **and**
>>> Was employed in agriculture fewer than 13 weeks in previous calendar year.
>>
>> H-2A non-immigrant agricultural workers.
>
> Employers must:
>
>> Require employees to complete an IRS W-4 "Employees Withholding Allowance Certificate"
>>> If employee wishes to claim full exemption from withholding on line 7 of the W-4 form, the employer may accept the claim of exempt status.
>>>
>>> If employee claims more than 10 exemptions or if a claim of full exemption is made and the employee earns more than $200 per week, the

> employer must send to INS copies of the W-4 form.

> Withhold income taxes on wages and remit to IRS through a local bank. The value of meals and lodging is generally not included in wages.

> Give employees Wages and Tax Statement (Form W-2) by January 31 for the preceding year and send Copy "A" of Form W-2 and Form W-3 together with Employers Annual Tax Return for Agricultural Employers (Form 943).

Employers may rely on a farm labor contractor for withholding income taxes from workers' wages. However, if the contractor fails to submit withheld taxes, the employer may be held liable for the tax and penalty under the joint employer principle.

The DOL Wage and Hour Division may enforce failure to deposit withheld F.I.C.A. and income taxes in a timely manner as a failure to pay *wages* when due, which would add DOL fines and penalties to the fines and penalties imposed by the IRS.

Migrant and Seasonal Agricultural Worker Protection Act (MSPA)[11]

The Migrant and Seasonal Agricultural Worker Protection Act (MSPA) ensures that migrant and seasonal farm workers receive information about and are protected by standards on pay rates, deductions, working conditions, housing, transportation and employment activities.

Exemptions: There are exemptions for certain family businesses, small businesses, common carriers, labor organizations, nonprofit charitable organizations, certain local short-term contracting activity, and certain custom poultry, custom combining, shade tobacco, hay harvesting, sheep shearing, and seed production operations.[12]

Definitions[13]

> An agricultural employer is any person who owns or operates a farm, ranch, processing establishment, cannery, gin, packing shed or nursery, or who produces or conditions seed, and who either recruits, solicits, hires, employs, furnishes, or transports any migrant or seasonal agricultural worker.

---

[11] Regulations Migrant and Seasonal Agricultural Worker Protection, Part 500, WH Publication 1455.

http://dol.gov/dol/esa/public/regs/cfr/29cfr/toc_Part500-899/0500.0075.htm

[12] http://dol.gov/dol/esa/public/regs/cfr/29cfr/toc_Part500-899/0500.0030.htm

[13] http://dol.gov/dol/esa/public/regs/cfr/29cfr/toc_Part500-899/0500.0020.htm

A migrant agricultural worker is a person engaged in agricultural employment of a seasonal or temporary nature and who is absent overnight from his or her permanent residence.

A seasonal agricultural worker is a person who is similarly employed but is not required to be away overnight from his or her permanent residence.

A farm labor contractor (FLC) is any person paid to recruit, solicit, hire, employ, furnish, or transport any migrant or seasonal agricultural worker.

An agricultural employer has practically the same obligations and liabilities as a farm labor contractor except for the registration requirements.

A farmer who engages a FLC must determine that the FLC possesses a current registration certificate authorizing the specific activities for which the FLC may be utilized. These include farm labor contracting, housing, transporting, and driving.

A farmer who engages a FLC is responsible for determining that the FLC possesses current registration certificates and vehicle insurance coverage for each activity, such as housing or transportation, the FLC is expected to perform.

A farmer who engages a FLC is, in most instances, deemed to be a joint employer with the FLC and, thus, liable for violations committed by that FLC.

The major requirements of MSPA are:

1. FLCs and each of their employees must possess a current certificate of registration from the U.S. Department of Labor. This certificate must specifically include all intended contracting activity including specific housing, transportation in specific vehicles, and driving if performed.

2. Employers and FLCs must at the time of recruitment disclose in writing, and in a language understood by the worker, to migrant and, if requested, to seasonal agricultural workers:

> Place of employment.
> Wage rates, including piece rates, to be paid.
> Crops and kinds of work.
> Period of employment.
> Transportation, housing, and any other benefits provided, including cost to the worker.
> Whether workers compensation is provided.
> > Name of carrier.
> > Name of policy holder.
> > Name and telephone number of each person who must be notified

of an injury or death and the time period during which the notice must be given.
Whether unemployment compensation is provided.
Whether a strike or work stoppage is in progress.
Any commission (kickback) arrangement between the employer and any local merchant.
Terms and conditions of housing, if any (see below).

3. Employers and FLCs must provide each pay period to each employee, and keep for three years, statements written in a language understood by the worker (optional form WH-501) that includes:[14]

Employee's name.
Employee's permanent address.
Employee's social security number.
Basis on which wages are paid.
Number of piecework units earned, if paid on a piecework basis.
Number of hours worked.
Total pay period earnings.
Specific sums withheld and the purpose of each sum withheld.
Net pay.
Employer's name.
Employer's address.
Employer's Internal Revenue Service identification number.

FLC's must also furnish wage records to each person for whom the FLC provides workers. Those who receive these records are required to keep them for 3 years.

4. If transportation is provided, vehicles used must be safe and, unless the owner possesses a liability bond approved by the Secretary of Labor, insured for liability not less than $100,000 for each seat in the vehicle, but not more than $5,000,000 for any one vehicle. An alternative procedure is workers compensation plus liability insurance for non-employee passengers and for periods when workers compensation coverage does not apply. In either case property liability insurance for at least $50,000 must be carried.[15]

If the vehicle is to be used for day-haul, and if the vehicle will transport migrant workers more than 75 miles round-trip, it is subject to Interstate Commerce Commission regulations. These include driver qualification and driving standards as well as vehicle standards.

---

[14] http://dol.gov/dol/esa/public/regs/cfr/29cfr/toc_Part500-899/0500.0080.htm

[15] http://dol.gov/dol/esa/public/regs/cfr/29cfr/toc_Part500-899/0500_Subpart_D_toc.htm

If the transportation is provided by a FLC, the FLC's transportation authorization certificate must be current and indicate the particular vehicle to be used.

Car Pooling. If car pooling is organized or paid for by the grower, the transportation and driver probably will be subject to MSPA registration and insurance requirements. If the car pooling is voluntary, but the driver collects an amount greater than the actual cost, the driver may be deemed to be a FLC–without the necessary certificate, medical examination, or insurance–and the grower may be utilizing an unregistered and out-of-compliance FLC.

Drivers must meet Department of Transportation physical requirements and have a medical examination and certification (Form WH-515) within the past 36 months. Proof of certification must be carried while driving.

Drivers must be at least 21 years of age with at least one year of driving experience.

Drivers must possess a valid driver's permit for the type of vehicle to be operated.

Drivers must be able to read and speak English sufficiently to understand traffic signals and to respond to official inquiries.

For 16 or more passengers (or other trucks with gross weight in excess of 26,000 pounds), a commercial driver's license is required and, with narrow exceptions, driver must be tested periodically for use of alcohol and controlled substances.[16]

5. If housing, other than public accommodations, is provided *by any person* it must meet local, state, DOL, and OSHA safety and health standards. In addition to posting a State certificate of occupancy, each employer, landlord, or FLC must post continuously, or provide to each worker at the time of recruitment, information on the terms and conditions of occupancy (Form WH-521) that includes:

>Name and address of the employer providing the housing.
>Name and address of the person in charge of the housing.
>Address and telephone number where occupants can be reached.
>Who may live in the housing.
>The charge (rent) to be made for the housing.
>Meals to be provided and the cost to the workers.
>Any charges for utilities.

---

[16] Controlled Substances and Alcohol Use and Testing, 49 C.F.R. Chapter III., Part 382, U.S. Department of Transportation, Federal Highway Administration, Bureau of Motor Carrier Safety.

Any other charges or conditions of occupancy.

6. The MSPA employee rights poster (DOL Form 1376) must be displayed at the workplace at all times.

Private Right of Action by Workers. Any person claiming a MSPA violation by an FLC, employer, or housing provider may be awarded actual damages or statutory damages of up to $500 per violation. If the MSPA violation caused death or bodily injury and there is workers compensation coverage, the employee may recover under workers compensation and sue also for MSPA statutory damages. Certain violations of the transportation provisions allow the possibility of statutory damages of $10,000 per violation. Full liability may be found against any joint employer.

Occupational Safety and Health Act (OSHA)[17]

Provides general safety and health standards for employers and employees.

Coverage: Agricultural employers engaged in interstate commerce who have a migrant labor camp or have employed 11 or more workers at any time during the previous 12 months.

Exempt: farm operators who employ only immediate family members. Exemption means exempt from audits and inspections but does not relieve the employer from the responsibility for maintaining a safe and healthy workplace.

Employers must:

> Furnish a workplace free from recognized hazards that could cause death or serious physical hazards. This requirement enables an OSHA inspector to cite an employer even if no specific standard exists.
>
> Inform employees of safety regulations.
>
> Instruct employees in the safe operation of tractors.
>
> Post OSHA Job Safety and Health Poster
>
> Report within 8 hours to the OSHA area office any fatal accident to an employee or any accident hospitalizing three or more employees.
>
> Maintain, within six working days, records of all occupational injuries and illnesses.

---

[17] http://hammock.ifas.ufl.edu/txt/fairs/19491

Post in a conspicuous place during the month of February the annual summary of your OSHA No. 200 log.

Retain all records of occupational injuries and illnesses for five years.

OSHA has the following standards that apply specifically to agriculture:
Temporary labor camps.[18]
Rollover protective structures.
Guarding of farm field equipment, farmstead equipment, and cotton gins.
Slow moving vehicle emblems, signs, and tags.
Storage and handling of anhydrous ammonia.[19]
Field sanitation[20] (enforced by Wage and Hour Division).

The Hazardous Communication Standard requires all covered employers to instruct employees in the safe handling of hazardous chemicals and make their written Hazard Communication Program available to employees, employee representatives, OSHA officials, and officials of the U.S. Department of Health and Human Services. HCS requires employers to:

Develop and implement a written Hazard Communication Program.

Provide to employees material safety data sheets (MSDSs), usually provided by suppliers.

Instruct employees how hazardous materials will be labeled and other warnings communicated.

Provide a list of hazardous chemicals present in the workplace.

Develop a method for informing employees of contractors of the hazards they may be exposed to in the workplace.[21]

---

[18]   29 C.F.R. 1910.142
        http://hammock.ifas.ufl.edu/txt/fairs/19511

[19]   http://hammock.ifas.ufl.edu/txt/fairs/19509

[20]   29 C.F.R. 1928.110
        http://hammock.ifas.ufl.edu/txt/fairs/19477

[21]   29 C.F.R. 1910.210
        Federal register Vol. 52, No. 163, August 24, 1987, Pages 31851-31886
        http://hammock.ifas.ufl.edu/txt/fairs/19494

Personal Responsibility and Work Opportunity Reconciliation Act of 1996

Requires employers to send within 20 days a copy of the IRS W-4 form on their newly-hired employees to the State Directory of New Hires. States will match New Hire reports against their child support records to locate parents, establish a child support order, or enforce an existing order. This information may also be used to prevent erroneous benefit payments under unemployment insurance, workers compensation, or public assistance programs. States will forward information to the National Directory of New Hires maintained by the Office of Child Support Enforcement of the U.S. Department of Health and Human Services.

State agency contact numbers:

Delaware	302-577-4815 X 249
Maryland	410-347-9911
New Jersey	609-588-2355
New York	800-972-1233, option 1
Pennsylvania	717-787-6466
Virginia	800-979-9014
West Virginia	800-835-4683

Plant Closing Notification/Layoffs - WARN Act

The WARN Act covers employers with 100 or more employees, excluding part-time employees or, 100 or more employees who, in the aggregate, work at least 4,000 hours per week.

Covered employers must notify workers of any intent to shut down work or discontinue employment for 50 or more workers at a single job site for a period of thirty days or more.

Polygraph Protection for Employees

Most private employers are prohibited from using lie detector tests either for pre-employment screening or during the course of employment.

All employers, even though they never require polygraph tests are required to display a poster (WH-Publication 1462) in a prominent place where all employees and job applicants can see it.

Social Security and Medicare Taxes[22]

Farm employers must make Social Security (6.2 percent) and Medicare (1.45 percent) deductions:

If they pay an employee $150 or more in wages during a calendar year.

If they pay total wages of $2,500 or more per year to all employees.

From holders of INS I-688A, I-688, and I-151 immigration documents. Employees in these immigration statuses are considered residents for purposes of Social Security and Medicare.

Employers may rely on a farm labor contractor to deduct Social Security and Medicare taxes from workers' earnings. However, if the contractor fails to submit withheld taxes, the employer may be held liable for the taxes and penalties under the joint employer principle.

Exemptions:

A child under 18 years of age in the employ of a parent.

Employers are not required to make deductions from earnings of H-2A workers because they are ineligible for benefits from these programs.

Social Security Number Verification

Employers may violate the law if their Social Security reports contain inaccurate information. An increasing number of employers report receiving letters from the Social Security Administration (SSA) warning of potential penalties to employers and employees if reported social Security numbers (SSNs) do not match SSA records. When a high percentage of SSNs are mismatched, SSA has rejected the tax reports and notified the employers that their reports must be corrected within 45 days. If migrant workers have moved on, correcting the reports may be impossible unless they were due to transcribing errors. In addition, immigration officials indicate that SSA is sharing mismatch data with the Immigration and Naturalization Service.

To help insure more accurate information, SSA has instituted the Enumeration Verification System (EVS) which allows an employer to call an 800 number to verify SSNs. Employers who wish to participate in EVS may send a letter or fax on company

---

[22] Circular A, Agricultural Employer's Tax Guide, Publication 51, Internal Revenue Service.

Farmer's Tax Guide, Publication 225, Internal Revenue Service.

letterhead requesting the use of EVS. Include your street address (no P.O. Box), your nine-digit Employer Identification Number, the name and telephone number of a contact person to:

> Social Security
> OCRO, Division of Operations Support
> 5-E-10 North Building
> 300 N. Greene Street
> Baltimore, Maryland 221201-1581

or FAX to 410-966-3366 or 410-966-9439.

If employers are informed that employees names and numbers are mismatched, SSN informs them that they must ask affected employees to show them their Social Security cards to ensure that the names and SSNs were reported accurately. If they were reported correctly, SSA directs the employer to have the employees call a toll free number to try to correct the problem. It is advisable for the employer to document the actions taken in this respect to establish a good-faith defense against potential penalties. Since the employer faces a potential penalty for failing to correct the numbers, it may be reasonable for an employer to discipline or terminate an employee who, after proper and sufficient notice, fails to correct the problem. If such steps are taken, it should be made clear that such action is being taken solely because of non-compliance with the legal obligation of both the employer and the employee to report accurate information to SSA. SSA states that anyone who obtains verification information for any purpose other than for tax filing purposes may be punished by a fine or imprisonment or both.[23]

If the employee succeeds in correcting a mismatched SSN, and that employee used a Social Security card in the I-9 employment eligibility verification process, the employer now has knowledge that the original I-9 verification was incorrect. However, the employer knows that the worker now possesses a valid card. It would be prudent to attach a copy of the correct card with the I-9 form and the copy of the original card.

Unemployment Compensation

This Federal/State program is intended to benefit persons unemployed through no fault of their own. For detailed information, contact your State industrial commission or unemployment insurance agency.[24]

---

[23] Federal Privacy Act, 5 USC 552a(I)

[24] http://www.usda.gov/agency/oce/oce/labor-affairs/statempl.htm
http://parker.itsc.state.md.us/

Worker Protection Standard (WPS) - Environmental Protection Agency[25]

Agricultural employers who use pesticides in crop production are required to inform, train, and protect workers from contact with pesticides.

Exemptions: There are exemptions from some WPS provisions for family members and crop advisors.

The principal WPS requirements include:

> The farm owner is generally responsible for all workers on the farm although certain responsibilities may be shared with farm labor contractors and EPA will determine liability on a case by case basis.
>
> Notification, warning signs, and/or posting to workers of pesticide applications as indicated on the pesticide label.
>
> Posting in a central location of specific information regarding pesticide use for 30 days after the restricted entry interval has expired. Posting must include:
>
>> Location and description of treated area
>> Date and time pesticide is to be applied
>> Product name, EPA registration number, and active ingredients.
>> Restricted entry interval for the pesticide
>> Name, address, and telephone number of emergency medical facility.
>> WPS safety poster.
>
> Each field worker must receive approved pesticide safety training at least once every five years. Employers may rely on current EPA training certificate as evidence of worker's training.
>
> Pesticide handlers are prohibited from applying pesticides in a manner that exposes workers or other persons.
>
> Workers must be excluded from areas being treated with pesticides until expiration of the restricted entry interval. (Some narrow exceptions exist.)
>
> Workers must be notified of areas to be treated so they can avoid exposure.

---

[25] "The Worker Protection Standard For Agricultural Pesticides: How to Comply: What Employers Need to Know," EPA 735-B-93-001, Environmental Protection Agency, July 1993.

http://www.usda.gov/agency/oce/oce/labor-affairs/wpspage.htm

Pesticide handlers must be trained, provided with appropriate personal protective equipment, and, in some instances, monitored while handling pesticides.

Decontamination supplies–soap, water, towels, eye-flush--must be provided.

Emergency assistance and transportation must be afforded exposed workers.

Workers must be provided information regarding pesticides to which they may have been exposed.

EPA Worker Protection Standard (CFR Title 40. Part 170)

David M. Scott
Pesticide Certification and Education Specialist
Pennsylvania Department of Agriculture
Bureau of Plant Industry
Division of Health and Safety

— David M. Scott's speaker biography appears on page 137 —

INTRODUCTION

EPA WORKER PROTECTION STANDARD
(CFR Title 40. Part 170)

The information presented here is to provide an overview of the Worker Protection Standard (WPS). The purpose of the rule is to reduce the long-term exposure of agricultural employees to pesticides and pesticide residues in their work place.

The WPS requires employers engaged in the production of an agronomic crop who use pesticides to provide a wide range of protection for their employees and others, including, education about pesticides, protection from pesticide exposure, decontamination and medical treatment of pesticides if exposure occurs.

The WPS was developed from six years of meetings and hearings. The final rule was published in the Federal register on August 21, 1992. The phase-in of the rule was completed on April 15, 1994. This rule requires employers of agricultural workers to provide a wide range of training, pesticide information, safety and decontamination supplies and medical treatment. It is important to remember that the WPS applies to the following:

(1) Those who own or manage a farm forest, nursery or greenhouse where pesticides

are used in production of agricultural plants.

(2) Those who hire or contract for services of agricultural workers (this includes labor contractors and others who contract with growers) to do tasks related to agricultural plants on farms, forest, nursery or greenhouse.

(3) You operate a business in which you (or people you employ) apply pesticides that are used for production of agricultural plants.

(4) You operate a business in which you (or people you employ) perform tasks as a crop advisor on a farm, forest, nursery or greenhouse.

With the exception:

(1) Those that employ only immediate family, specifically: spouse, children, step-children, foster children, parents, brothers and sisters for work on their own farms, forest, nursery and greenhouse.

The standard divides employees into two types: Workers and Pesticide Handlers
The term under which the employee falls depends on the type of task being performed by the employee at the time,

Workers

A worker is anyone who:

(1) Is employed (including self-employed) for any type of compensation and

(2) Are doing tasks, such as harvesting, weeding or watering, relating to the production of agricultural plants on a farm, forest, nursery or greenhouse. This term does NOT include persons who are employed by a commercial establishment to perform tasks as crop advisors.

Pesticide Handlers:

A pesticide handler is anyone who:

(1) Is employed (including self-employed) for any type of compensation by an agricultural establishment or a commercial pesticide handling establishment that uses pesticides in the production of agricultural plants on a farm, forest, nursery or greenhouse and,

(2) Are doing any of the following tasks:

> mixing, loading, transferring or applying pesticides,
>
> handling opened containers of pesticides,
>
> acting as a flagger,
>
> cleaning, handling, adjusting or repairing the parts of mixing, loading or application equipment that may contain pesticide residues,
>
> assisting with the application of pesticides, including incorporating the pesticide into the soil after the application has occurred,
>
> entering a greenhouse or other enclosed area after application and before the inhalation exposure level listed on the product labeling has been reached or one of the WPS ventilation criteria has been met to:
>
>> operate ventilation equipment,
>>
>> adjust or remove coverings, such as tarps, used in fumigation or,
>>
>> check air concentration levels,
>
> entering a treated area outdoors after application of any soil fumigant to adjust or remove soil coverings, such as tarpaulins,
>
> performing tasks as a crop advisor:
>
>> during any pesticide application,
>>
>> before any inhalation exposure level or ventilation criteria listed in the labeling has been reached or one of the WPS ventilation criteria has been met,
>>
>> during any restricted-entry interval,
>
> disposing of pesticides or pesticide containers.

A person is NOT a handler if he or she only handles pesticide containers that have been emptied or cleaned according to instructions on pesticide product labeling or, if the labeling has no such instructions, have been triple-rinsed or cleaned by an equivalent method, such as pressure rinsing.

A person is NOT a handler if he or she:

(1) Is only handling pesticide containers that are unopened AND

(2) Is not, at the same time, also doing any handling task (such as mixing or loading).

Depending on the type of agricultural production you are involved with, the WPS provides for many, exceptions that may permit you to do less or provide options that may involve different requirements. These exceptions and options are described in the "Worker Protection Standard for Agricultural Pesticides, How to Comply Manual", developed by the EPA to assist employers with compliance. The following list of requirements represents the **Maximum** WPS requirements. You will be in compliance with WPS if the following requirements are met.

DUTIES FOR ALL EMPLOYERS

Anti-Retaliation

Do not retaliate against a worker or handler who attempts to comply with the WPS.

Information at a Central Location

(1) In an easily seen central location on each agricultural establishment, display close together:

EPA WPS safety poster,

name, address and telephone number of the nearest emergency medical facility,

these facts about each pesticide application (from before each application begins until 30 days after the restricted-entry interval REI.

Product name, EPA registration number and active ingredient(s),

location and description of treated area,

time and date of application, and REI.

(2) Tell workers and handlers where the information is posted and allow them access.

(3) Tell them if emergency facility information changes and update the posted information.

(4) Keep the posted information legible.

Pesticide Safety Training

Unless they possess a valid EPA-approved training card, train **handlers and workers** before they begin work and at least once every five (5) years:

> use written and/or audiovisual materials,
>
> use EPA WPS worker training materials for training handlers,
>
> use EPA WPS worker training materials for training workers,
>
> have a certified applicator conduct the training orally and/or audiovisually in a manner the employees can understand, using easily understood terms and respond to questions.

Decontamination Sites

(1) Establish a decontamination site with ¼ mile of all workers and handlers. Supply:

> enough water for routine and emergency whole-body washing and for eyeflushing,
>
> plenty of soap and single-use towels,
>
> a clean coverall.

(2) Provide water that is safe and cool enough for washing, for eyeflushing and for drinking. Do not use tank-stored water that is also used for mixing pesticides.

(3) Provide **handlers** the same supplies where personal protective equipment (PPE) is removed at the end of a task.

(4) Provide the same supplies at each mixing and loading site.

(5) Make at least one pint of eyeflush water immediately accessible to each **handler**.

(6) Do not put **worker** decontamination sites in areas being treated or under an REI.

(7) In areas being treated, put decontamination supplies for **handlers** in enclosed containers.

Employer Information Exchange

(1) Before any application, commercial handler employers must make sure the operator of the agricultural establishment, where a pesticide will be applied is aware of:

> location and description of area to be treated,
>
> time and date of application,
>
> product name, EPA registration number, active ingredient(s) and REI,
>
> whether the product label requires both oral warnings and treated area posting,
>
> all other safety requirements on labeling for workers or other people,

(2) Operators of agricultural establishments must make sure any commercial pesticide establishment operator they hire is aware of:

> specific location and description of all areas on the agricultural establishment where pesticides will be applied or where an REI will be in effect while the commercial handler is on the establishment,
>
> restrictions on entering those areas.

Emergency Assistance

When any handler or worker may have been poisoned or injured by pesticides:

(1) Promptly make transportation available to an appropriate medical facility.

(2) Promptly provide to the victim and to medical personnel:

> product name, EPA registration number and active ingredient(s),

all first aid and medical information from label,

description of how the pesticide was used,

information about victim's exposure.

ADDITIONAL DUTIES FOR WORKER EMPLOYERS

Restrictions During Applications

(1) In areas being treated with pesticides, allow entry only to appropriately trained and equipped handlers.

(2) Keep nursery workers at least 100 feet away from nursery areas being treated.

(3) Allow only handlers to be in a greenhouse:

during a pesticide application,

until labeling-listed air concentration level is met or, if no such level, until after two (2) hours of ventilation with fans.

Restricted-Entry Intervals (REI's)

During any REI, do not allow **workers** to enter a treated area and contact anything treated with the pesticide to which the REI applies.

Notice About Applications

(1) Orally warn workers **and** post treated areas if the pesticide labeling requires.

(2) Otherwise, **either** orally warn workers or post entrances to treated areas. Tell workers which method is in effect.

(3) Post all greenhouse applications.

Posted Warning Signs:

(1) Post legible 14" X 16" WPS-design signs just before application; keep posted during REI; remove before workers enter and within three (3) days after the end of the REI.

(2) Post signs so they can be seen at all entrances to treated areas, including entrances from labor camps.

Oral Warnings:

(1) Before each application, tell workers who are on the establishment (in a manner they can understand):

location and description of treated area,

REI, and not enter during REI.

(2) Workers who enter the establishment after application starts must receive the same warning at the start of their work period.

ADDITIONAL DUTIES FOR HANDLER EMPLOYERS

Application Restrictions and Monitoring

(1) Do not allow handlers to apply a pesticide so that it contacts, directly or through drift, anyone other than trained and PPE-equipped handlers.

(2) Make sight or voice contact at least ever two (2) hours with anyone handling pesticides labeled with skull and crossbones.

(3) Make sure a trained handler equipped with labeling-specified PPE maintains constant voice or visual contact with any handler in a greenhouse who is doing fumigant-related tasks, such as application or air-level monitoring.

Specific Instructions for Handlers

(1) Before handlers do any handling task, inform them, in a manner they can understand, of all pesticide labeling instructions for safe use.

(2) Keep pesticide labeling accessible to each handler during entire handling task.

(3) Before handlers use any assigned handling equipment, tell them how to use it safely.

(4) When commercial handlers will be on an agricultural establishment, inform them beforehand of:

areas on the establishment where pesticides will be applied or where an REI will be in affect,

restrictions on entering those areas,

(The agricultural establishment operator must give you these facts.)

Equipment Safety

(1) Inspect pesticide handling equipment before each use and repair or replace as needed.

(2) Allow only appropriately trained and equipped handlers to repair, clean or adjust pesticide equipment that contains pesticides or residues.

Personal Protective Equipment (PPE)

Duties Related to PPE:

(1) Provide handlers with the PPE the pesticide labeling requires for the task and be sure it is:

clean and in operating condition,

worn and used correctly,

inspected before each day of use,

repaired or replaced as needed.

(2) Be sure respirators fit correctly.

(3) Take steps to avoid heat illness.

(4) Provide handlers a pesticide-free area for:

storing personal clothing not in use,

putting on PPE at start of task,

taking off PPE at end of task.

(5) Do not allow used PPE to be worn home or taken home.

Care of PPE:

(1) Store and wash used PPE separately from other clothing and laundry.

(2) If PPE will be reused, clean it before each day of reuse, according to the instructions from the PPE manufacturer unless the pesticide labeling specifies other requirements. If there are no other instructions, wash in detergent and hot water.

(3) Dry the clean PPE before storing, or hand to dry.

(4) Store clean PPE away from other clothing and way from pesticide areas.

Replacing Respirator Purifying Elements:

(1) Replace dust/mist filters:

when breathing becomes difficult,

when filter is damaged or torn,

when respirator label or pesticide label requires (whichever is shorter), **OR**

at the end of day's work period, in the absence of any other instructions or indications.

Disposal of PPE:

(1) Discard coveralls and other absorbent materials that are heavily contaminated with undiluted pesticide having a DANGER or WARNING signal word.

(2) Follow Federal, State and local laws when disposing of PPE that cannot be cleaned correctly.

Instructions for People Who Clean PPE:

Inform people who clean or launder PPE:

that PPE may be contaminated with pesticides,

> of the potentially harmful effects of exposure to pesticides,
>
> how to protect themselves when handling PPE,
>
> how to clean PPE correctly.

Handlers who are certified pesticide applicators must be given all of the WPS handler protections, except that they need not receive WPS training.

Areas of note:

The WPS does not cover animals or workers making applications of pesticides made to animals or structures housing them. Nor does it apply to lawn care, golf course, right-of-way or other non-plant production applications.

WPS covers only workers and handlers, not customers, even those at pick-your-own operations.

Retail sales areas, separated from production areas are also exempt from the rule, however the entire production facility cannot be exempted as a sales area.

For more information, contact your Regional EPA office or state agency for information.

EPA REGIONAL OFFICES:

Region 1 (MA, CT, RI, NH, VT, ME)
U.S. Environmental Protection Agency, Region 1
Pesticides and Toxic Substances Branch (APT)
1 Congress St.
Boston, MA 02203
(617) 565-3273

Region 2 (NY, NJ, PR, VI)
U.S. Environmental Protection Agency, Region 2
Pesticides and Toxic Substances Branch (MS-105)
2890 Woodridge Avenue, Building #10
Edison, NJ 08837-3679
(212) 637-3000

Region 3 (PA, MD, VA, WV, DE)
U. S. Environmental Protection Agency, Region 3
Toxics and Pesticides Branch (3AT-30)
841 Chestnut Building
Philadelphia, PA 19107
(215) 814-2128

The rule is also available on the World Wide Web @ www.access.gpo.gov/nara/cfr

(ref) 40 cfr.170

Worker Protection Standard for Agricultural Pesticides
How to Comply (141 pages) is available from:

U.S. Government Printing Office
Superintendent of Documents, Mail Stop: SSOP
Washington, DC 20402-9328
    Order #: ISBN – 0-16-04-1939-5

Gempler's
PO Box 270
211 Blue Mounds Road
Mt. Horeb, WI 53572
1-800-551-1128
1-800-382-8473
    Order code: 112A
    Order # : EPA-HTC or H-W-40

Developing a Safety Training Program Involves a Commitment to Reduce Hazards and Injuries

Ellen L. Abend, R.N., M.S.
Health Educator
Department of Population Medicine
and Diagnostic Sciences
Cornell University

Eric M. Hallman, M.S.
Director
Agricultural Health and Safety
Department of Population Medicine
and Diagnostic Sciences
Cornell University

INTRODUCTION

Farms in the Northeast region are growing in terms of number of acres, total investment, production per farm, and labor force size. The need for an increase in employees has forced a shift from traditional family-only labor to the employment of non-related workers. According to the National Agricultural Statistics Service (NY Ag. Statistics, 1997) there were approximately 61,300 people working in agriculture in New York during 1997. If unpaid (family) labor is excluded, 82% of New York's agricultural work force is covered by state-mandated workers' compensation insurance. However, the New York Workers' Compensation Insurance program does not provide a direct payment incentive to small businesses, such as agriculture, to control injuries in the work place. As a result, the majority of agricultural employees receive no formal training in occupational safety and health issues related to injury prevention and wellness.

The goal of the NIOSH funded New York State Agricultural Hazard Abatement and Training (AHAT) project was to develop an agricultural injury prevention model which would demonstrate that farm owners would voluntarily take positive steps toward abating on-farm hazards and reducing workers' injuries through improvement in safe work practices if a meaningful incentive was available. Initially, the incentive offered in the AHAT project was a direct rebate of up to 10% of a farm's annual workers' compensation insurance premium. The rebate was performance-based which provided an opportunity for researchers to observe how much effort farm owners were willing to put into improving safety on their farms.

BACKGROUND

Hazards are Prevalent on Farms

A hazard, as defined by the American Society of Safety Engineers, is "a condition or changing set of circumstances that presents a potential for injury, illness, or property damage. The potential or inherent characteristics of an activity, condition or circumstance which can produce adverse and harmful consequence" (Murphy, 1992 p.15). There are a great number of hazards that can be observed in production agriculture due to the diversity of tasks required to produce food or fiber commodities. While agricultural hazards overlap into different categories depending on the circumstances, Murphy (1992) suggests that, generally, they are associated with tractors, machinery, animals, chemicals, toxic gases, dusts, noise, and weather.

From 1993-1996 on-site hazard audits were conducted on 580 farms located in four multi-county regions within New York State. These safety audits were part of the NIOSH sponsored Farm Family Health and Hazard Surveillance (FFHHS) project. Trained professional safety and health personnel evaluated many aspects of each farm from a safety perspective including: tractors; equipment; structures; waste handling/storage; electrical components; chemical storage; and general farmyard layout. This type of surveillance gave an accurate picture of hazards currently present on farms. In focusing just on the tractors and equipment component of the survey, the following hazards were observed (Hallman, et al, 1997).

Of the 2,513 tractors inspected:
- 61.4% did not have a roll over protective structure (ROPS)
- 45% were missing the power take off (PTO) master shield
- 79.6% were not properly protected with slow moving vehicle (SMV) emblems

Of the 4,423 trailed equipment pieces (2,803 pieces PTO powered) inspected:
- 34.6% did not have the PTO shaft properly shielded
- 19.2% were missing the required guards over hazard areas
- 38.4% were not properly protected with SMV emblems

Often, farmers' perceptions of on-farm hazards differ from the ones found to be most commonly associated with injuries and fatalities. Rosenblatt and Lasley (1991) found that farmers perceived chemicals (herbicides/insecticides) as more hazardous than tractors, yet 50% of fatalities on farms involved tractor-related operations. This clearly suggests a misperception by farm owners and their employees about the day-to-day risks associated with working on farms.

Denis (1988) also suggested that farmers may impose hazards upon themselves and their workers due to lack of adequate information and formal training in machinery operation and animal handling. In blaming farmers for the lack of adequate worker training, it is assumed that there is relevant training readily available in the rural community. In reality, this is not the case. Denis further noted that, historically, there has been a lack of concern for occupational health and safety issues affecting the farm community and a lack of support for training by the institutions that farmers regularly interact with such as the government, farm organizations, agribusiness, and the medical professions. This lack of support, coupled with social and economic pressures to increase production and profits, may contribute to farmers exposing themselves and their workers to increased hazards and risk-taking activities.

Injuries and Fatalities on Farms

Although there is no one central agency that collects agricultural injury and illness statistics (Murphy, 1992), the National Safety Council (NSC) is considered a reliable source of work-related death and injury data. In its most recent publication the NSC estimates the fatality rate for agriculture to be 21/100,000 workers, which places it second only to mining which is designated as the most hazardous industry (NSC, 1997).

The Traumatic Injury Surveillance of Farmers (TISF) project (Meyers, 1997) estimated that in 1993, 201,081 lost-time work injuries occurred nationally. The leading "causes" of injuries were livestock, machinery, and hand tools. The TISF statistics also indicated that the dairy industry had the 2nd highest rate of lost-time injuries with the rate for workers (7.8) being higher than the rate for employers (5.7). Since dairy farms comprise the largest single farm-type of New York farms, the high rate of injuries to employees is an important concern.

New York State's farm-related injury and fatality picture mirrors that of the national breakdown. Pollock (1990) found that a large number of non-fatal injuries occurring to dairy employees predominantly involved two types of work: handling animals and repairing machinery. Twenty-one per cent of the reported injuries were animal-related and 19% involved machinery.

The high injury rate in agriculture involves several factors: long hours, solitary work, diversity of tasks, fatigue, older machinery, and lack of training (Stueland, Lee, & Layde, 1991). With the current adverse economic conditions in the farm community, some of the above factors cannot be readily addressed. However, the AHAT project developers felt that training employees to improve safe work practices was feasible and an important step toward reducing injuries to farm personnel.

Reducing Farm Injuries and Fatalities

Efforts to reduce injuries on farms generally involves education in safe work practices through modifying employee behaviors, implementing engineering controls, and enacting legislation to set standards (Aherin, Murphy & Westaby 1992). Using new technology to improve the design of equipment has been one method used for reducing injuries in agriculture, but it has not been entirely successful. In fact, Poppendorf & Donham (1991) indicate that increase mechanization in agriculture has led to increased hazards to which farm workers are exposed.

Goldenhar and Schulte (1994) suggest that implementing engineering controls without worker involvement or focusing only on behavior changes will not send a positive message to employees. Employees need to be empowered to feel they have a sense of personal control over their safety in the work place. Sorensen, et al (1996) found through a two-year WellWorks Project, that employees who perceived their employers as being willing to make work-site changes were more positive towards making individual health behavior changes. Two other studies conducted by Bayer (1984) and Becker and Shoup (1986) concluded that safety programs on large farms which involved management commitment, employee training, and incentives resulted in lower injury rates and increased production. All of these studies support the concept that employer involvement along with education of workers will promote changes in behaviors.

METHODS

Sample

Farms invited to participate in the AHAT project were randomly selected from 277 dairy farms that had previously participated in an on-farm safety audit which was part of the NIOSH-sponsored Farm Family Health and Hazard Surveillance (FFHHS) research effort. The sample was restricted to dairy farms because: (1) dairying is not a commodity area that lends itself readily to part-time enterprise; (2) dependency on hired labor is much greater than on other farm types; (3) and, according to New York State law, workers' compensation insurance must be provided for virtually all employees. Sample farms, therefore, had to meet three criteria:

- Be an active dairy farm.
- Have full or part-time employees.
- Provide workers' compensation insurance.

A recruitment letter was mailed to potential participants which included a pre-addressed, stamped return postcard on which they could respond 'yes' or 'no' or request more information before making a decision. One month after the letter was mailed, three attempts were made to contact non-responders by telephone.

Intervention

The intervention consisted of two components: (1) correcting hazards specific to guarding/shielding problems on equipment that were identified during an on-site safety audit; (2) establishing an on-going safety training program for employees.

Hazard Abatement

One of the outcomes of interest to this research project was observing what farmers would *voluntarily* choose to do when offered a performance-based incentive payment. In keeping with this concept, engineering recommendations unique to their farm, were mailed to each farm owner early in the project timetable. The farmers were asked to choose at least five specific hazards to correct from an average list of eleven. These recommendations were based on hazard data obtained on each individual farm during the FFHHS on-site visit. Farmers were asked to note the corrections made on a "Hazard Abatement Record" (HAR) and return the form by the end of the project.

Farm Worker Training Manual

The AHAT project staff developed a resource guide for farm owners to use as they worked toward establishing an ongoing safety program on their farms which included training employees in agricultural health and safety issues. This guide contained the materials and concepts covered in the training seminars plus additional information which employers would need on their "journey into the worker training arena." After peer review for relevancy, the resource guide was presented to a focus group of New York farm owners for input as to whether it was user-friendly. The section of the guide that focused on establishing an on-going safety and health program was piloted with two dairy farm manager groups.

This resource entitled, *Farm Worker Training Manual,* was divided into four sections:

- Developing a Safety & Health Program
  This guides the user through a four-step process to aid in developing a farm's safety program involving: identifying safety and health problems on the farm; researching the problems; taking measures to reduce risks and hazards; and developing an effective employee training program.
- Regulatory Compliance
  An explanation of various laws, regulations, and compliance requirements that relate to agricultural safety and health.
- Reference Materials
  Actual resource materials that are grouped by nine specific subject areas which provide background information for training preparation are included in this section.
- Farm Safety & Health Videos
  An extensive list of videotape training aids is provided.

Train-the-Trainer Seminars

Farm owners or their designated safety officers were required to attend two training seminars, two hours in length, which were held approximately three weeks apart in a local extension or farm credit office. Five separate geographical locations were chosen to reduce travel time for participants and facilitate attendance. To encourage participation, the seminars were held in the middle of the day and during the winter months, so as not to interfere with daily chores or the growing season. The goal of the seminars was to empower farm owners/safety officers to be able to establish safety programs and conduct employee training on their own farms.

Information about establishing an on-farm safety program was presented and each farm owner was given a *Farm Worker Training Manual.* Careful orientation to the manual was provided to show farmers how it could be used to help them establish their own safety program and perhaps deter them from just shelving the book.

As part of the empowerment process, the participants were introduced to three different methods of conducting trainings through the use of training guides provided in the *Farm Worker Training Manual*: On-site; Video based; and Formal. A "homework" assignment was given to each participant: conduct one training with employees before the second seminar. Participants were also given a 'Workers' Compensation History Form' to complete which provided information on number of employees, insurance carrier, 1996 premium payment, and number of claims filed in 1996.

During both seminars demonstrations of different training methods were presented. It was felt that the more exposure the participants had to different training methods, the more comfortable they would feel about developing their own training style. However, the main focus of the second seminar consisted of a group-sharing of the successes and challenges each individual faced when conducting the trial employee training on their individual farms. At the end of this seminar, each participant was asked to conduct safety training with his/her employees for the next consecutive six months and fill out a monthly report on activities conducted.

Rebate Formula

The amount of rebate was determined by the following formula based on a point system developed by the AHAT staff.

- Participation in training session I..............................5 points
- Participation in training session II.............................5 points
- Six monthly reports returned @ ½ pt. each....................3 points
- "Hazard Abatement Record" returned.........................1 point

Initially, full participation in the project was valued at a 10% rebate payment on a workers' compensation insurance premium paid. However, as the project continued, it was determined that developing a point system would address the levels of participation in a more equitable way. Since time commitment to attend the seminars was an important issue to participants *and* to those who chose not to participate in the AHAT project, the staff felt that attendance at these was worth more points than any other activity. If a participant completed all of the above activities, the farm would be eligible for a 14% rebate of the 1996 workers' compensation insurance premium paid.

RESULTS

Study Sample

Of the initial sample of 277 dairy farms that participated in the FFHHS hazard audit, 50 were placed in a control group which would receive no intervention and the remaining 227 were sent an offer letter explaining the project. Sixty-two farm owners indicated they were not eligible to participate because they either were no longer farming (6.5%) or did not employ any outside labor (93.5%). Sixty-seven farms declined to participate or could not be contacted via a follow-up telephone call. One farmer selected to receive an offer letter died from a farm injury just prior to the mailing. Ninety-seven farm owners agreed to participate in the project and 68 actually attended the training seminars which were required to be fully involved in the project. Of the 68 participating farms, three withdrew from AHAT within the first or second month, citing time constraints and loss of employees as reasons for withdrawal.

The participating farms were located in four specific geographical regions in the state which corresponded with the regions chosen for the FFHHS project. This regionalization allowed seminars to be conducted locally and increased the number of participating farms. Since time for dairy farmers to be away from their farms is limited, holding the seminars within a thirty minute drive from home was important. In May, 1997 a survey was mailed to the participants in the AHAT project *and* to those eligible farms who decided not to participate or did not respond to the offer letter. Seventy-four percent (48) of the AHAT project participants responded and 90% indicated that the location of the training seminar was an important factor in helping them decide to participate.

Ninety-four percent of the 65 participating farms completed a Workers' Compensation History form on which they indicated number of employees. There were a total of 575 employees reported for the 61 farms with a mean of 9.4 workers per farm. The number of employees per farm ranged from one employee on 11 farms to one farm with 101 employees. Figure 1 illustrates the distribution of the number of employees per farm. Using the standard OSHA designation of small (10 or less employees) versus large (11+ employees) farms, 78.7% (48) of farms had ten or fewer employees with a mean of 4.1 per farm.

Figure 1: Distribution of Employees of Participating Farms (n=61)

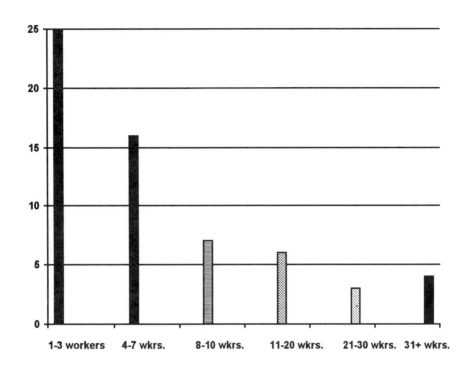

Hazard Abatement Summary

Participants were asked to *voluntarily* choose to correct some of the hazards pertaining to guarding/shielding aspects of equipment that had been identified during a previous on-site safety audit. A total of 722 hazards were identified with a mean of 11.1 per farm and these included:

- 135 missing/damaged PTO master shields on tractors
- 101 missing/damaged PTO shaft shields on equipment
- 79 missing/damaged guards/shields on equipment
- 370 missing/damaged SMV emblems on tractors or equipment
- 37 miscellaneous hazards such as water covers missing on electrical outlets in milk houses and guards missing from manure handling or fuel tank areas.

Since one of the outcomes of this research project was to determine what participants would *voluntarily* do if offered a performanced-based incentive, farm owners were encouraged, but not required to correct the hazards and their rebate payment reflected whether or not the Hazard Abatement Record (HAR) was returned, not how many hazards were corrected. Of interest was which hazards farm owners chose to correct. Of the 65 participating farms, 75.4% (49) returned the HAR indicating what engineering recommendations were followed (see Table 1).

The HAR also allowed for the participants to list the amount of money and time spent on correcting each hazard. Thirty-six farm owners chose to report dollars spent to correct hazards. The total amount spent by the 36 farms was $5,439.53 with a mean of $151.10 per farm. Thirty-eight farms chose to report the number of hours spent correcting hazards. The total number of hours reported was 183.95 with a mean of 4.8 hours per farm (Table 1).

Table 1: Hazard Abatement Activities in Terms of Time and Expenditures (n=65)

REGION	TOTAL # OF FARMS	# HAZARD RECORDS FILED	AVERAGE HRS. /FARM (N=36)	AVERAGE $/ FARM (N=38)
1	26	18 (69%)	5.4	$114.73
2	10	10 (100%)	4.1	$145.11
3	15	10 (67%)	6.6	$256.45
4	14	11 (79%)	3.3	$127.79
All Farms	65	49 (75%)	4.8 (36 farms)	$151.10 (38 farms)

Figure 2 indicates what type of equipment engineering corrections were made on the 49 farms that completed a HAR in relation to the number of hazards that were identified on those farms during the FFHHS safety audit. Only hazards that directly relate to guards and shields on equipment or tractors are addressed on the graph.

In looking at the types of hazards identified and the relative cost of individual corrections, it is interesting to note the following percentages:

- 80% of the PTO master shield hazards were addressed;
- 77.9% of the PTO shaft shield problems were corrected;
- 95% of the missing guards on equipment were corrected;
- 60.4% of missing or damaged SMV emblems were addressed.

The average PTO shaft shield costs $40-$50, however, the average SMV emblem costs $5.00 - $7.00. No research was conducted as to why the replacement of missing guards and PTO shields was at such a higher level than replacement of SMV emblems.

Figure 2: Number of Equipment Hazards Identified vs. Number Corrected on Farms Returning the Hazard Abatement Record (n=49).

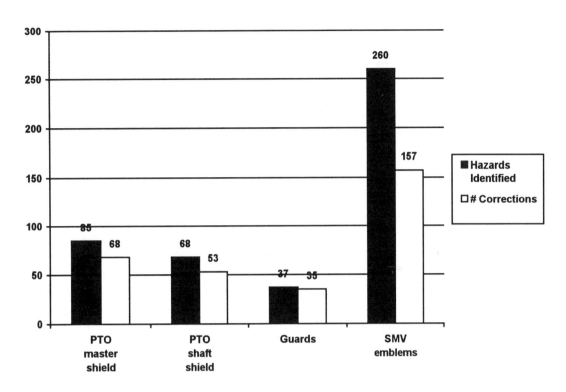

Further analysis of the hazard abatement activities indicated that some of the equipment was sold or put out of service on individual farms which would mean that the hazard was no longer present on that specific farm. Table 2 compares the number of equipment hazards repaired to pieces of equipment sold or no longer used.

Table 2: Type of Hazard Abatement Used on Tractors/Equipment by Farms Returning the HAR. (n=49)

Equipment	Total #	# Hazards Id.	# Repair/Replaced	# Sold	# Not Used
Tractors with master shield	280	85	55	11	2
Equip. w/PTO	289	68	39	8	6
Equip. w/guards	289	52	20	5	10
Tractors/equip. with SMVs	280/404 (684 total)	260	147	8	2

Finally, in order to document that hazard corrections were completed as stated on the HAR, on-site farm visits were made to 25% (16) of the participating farms. A cross-section of farms was chosen to include five farms with 11 or more employees and 11 farms who employed 10 or fewer employees which was similar in ratio to the number of participating farms whose employees were less than or more than eleven.

A total of 104 hazards were identified on the sixteen farms that were visited by AHAT personnel. Ninety-five percent (99) of the hazards had been addressed either through replacing guards, retiring, or selling equipment which is shown in Table 3.

Table 3: On-site Farm Hazard Correction Verification Visits (n=16)

HAZARD	# ID	# REPLACED	# SOLD	# NOT USED	%
Master shield missing	29	21	7	0	96.6%
PTO shield missing	19	10	6	2	94.7%
Equip. guards missing	10	7	2	2	90%
SMV emblem missing	38	31	4	1	94.7%
Miscellaneous	8	8	NA	NA	100%
Total hazards	104	77	19	5	95.2%

Implementing Employee Safety Training

Farm owners or their designated safety officers from 70.1% (68) of the 97 farms who agreed to participate in the AHAT project attended the training seminars which were held either in local cooperative extension or farm credit offices because they were familiar surroundings. These 68 farms employed a total of 575 workers.

Participants were encouraged, but not required to return a report each month that summarized that month's employee training activities. The following graph shows a comparison between the number of farms submitting reports and the number of farms which reported that employee training was conducted during a particular month (see Figure 3). As is noted on the graph, monthly report submitting during the total training period ranged from a high of 100% during January, 1997 to a low of 64% in September, 1997 which was the last month of the final group's reporting period.

There were 372 trainings conducted during the total training period (January - September,1997) as reported by the 61 farms which chose to submit one or more monthly reports. This resulted in a mean of 6.1 trainings conducted per farm during <u>each farm's six-month training period</u>. Two farms submitted reports stating that no trainings had been conducted during the six month training period. Four farms involved in the study did not submit any monthly training reports. Interestingly, three of these farms had previously submitted a workers' compensation history form which made them eligible for a rebate.

When comparing large farm (11+ employees) to small farm (<11 employees) participation, it was found that the large farms conducted an average of 8.1 trainings during the six month period, while small farms conducted an average of 6.1 trainings.

Figure 3: Comparing Percentage of Farms that Submitted Monthly Reports to Farms Actually Conducting Employee Safety Training. (n=61)

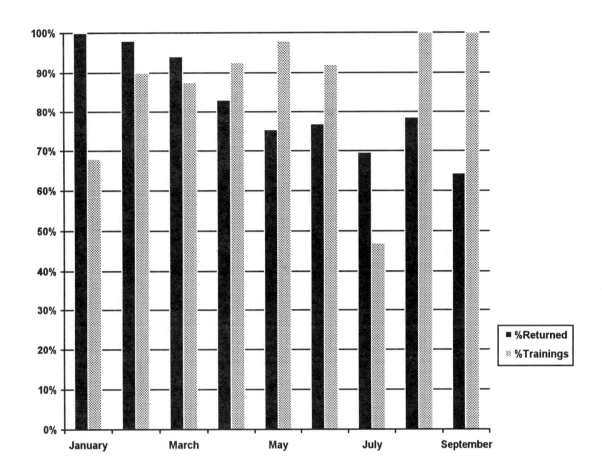

The participants were asked to identify the type (on-site, video-based, formal) of trainings conducted as well as the subject area covered. Of the 372 reported trainings: 73% (271) were identified as the on-site type; 21% (79) as video-based; and 6% (22) were formal (see Figure 4). It is interesting to note that the trainings identified as formal, were generally conducted by an invited guest or employees traveled off the farm to a safety meeting scheduled by an insurance company or implement dealer.

Since there was no requirement for a farm to conduct only one type of training, seven combinations of training types were possible. Forty-four percent (26) of the participants used only the on-site method while 30.5% (18) used a combination of on-site and video methods. Fifteen percent (9) used all three training types during the six month period but no participant used only the formal method.

Figure 4: Comparison of Types of Training Conducted by Farm Owners. (n=372)

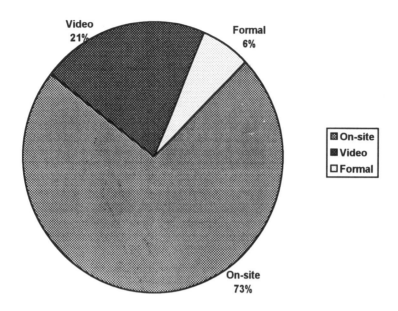

Training guides for 19 different safety topics were available in the *Farm Worker Training Manual* for use in employee training, and during the second seminar, the video-based training that was demonstrated was about animal handling. However, it was emphasized several times during the seminars and in mailed update information that the trainings chosen by each participant should be relevant to their own farm's needs. After all monthly reports had been received, the safety topics that were addressed were grouped into fifteen different categories. Table 4 lists the topic covered, number of times, and number of farms that addressed it.

Table 4: Safety Issues Addressed During Employee Training Sessions (n=372)

SAFETY ISSUE	# SESSIONS (N=372)	% TOTAL	# FARMS	% (N=61)
Animal Handling	74	19.9%	43	70.5%
Tractor Operation	45	12.1%	33	54%
Chemical Handling	33	8.9%	22	36%
PPE Use	31	8.3%	23	38%
Maintain Equipment	24	6.4%	20	33%
Skid Steer Operation	23	6.2%	19	31%
Hay Harvesting	23	6.2%	18	29.5%
Silos/Feed Systems	23	6.2%	16	26%
General Farm Safety	22	5.9%	18	29.5%
Shop/Wood Lot	18	4.8%	16	26%
PTO (specific)	13	3.5%	11	18%
Fire Safety	12	3.2%	11	18%
Manure Handling	11	3.0%	9	15%
Personal/Family	11	3.0%	7	11%
Back/Lifting	9	2.4%	8	13%

It should be noted that the participating trainers were not asked to indicate whether a training guide was used to prepare for employee training sessions. Also, it is interesting that for 67% (10) of the safety topics addressed during the employee training sessions training guides were available in the *Farm Worker Training Manual*. This suggests that the trainers liked to have a basis on which to build their own training program.

Finally, the monthly report form provided space for the project participants to comment about voluntary improvements made at the work site and observed changes in employees' work practices. A total of 393 written comments were recorded by 97% (59) of the 61 farms that submitted monthly reports:

- 17% addressed improvements to the work site, in addition to those involved in the hazard abatement component.
- 17% noted positive changes in attitude and safety awareness of employees.
- 15.6% directly referred to using PPE.
- 15% were related to equipment repairs or operator's behavior.
- 15% noted changes in employee behaviors as observed by the employers.
- 12% referred directly to working with livestock.

Some of the comments concerning observed employee behaviors that were *voluntarily* made by the farm owners were as follows:
- "The men...not as noisy around cattle."
- "All persons entering silos now unplug the cord...no one enters without a buddy."
- "Everyone, even me, uses seat belts now."
- "Traveling with bucket (bobcat) in down position."
- " Children are kept off all equipment..."
- "Does not put head under cows when reaching in for milking."
- "Employees no longer smoke inside farm buildings..."

As mentioned above, comments were also *voluntarily* made about improvements to the work site by farm owners. These improvements were in addition to the written engineering recommendations that were part of the hazard abatement component of the project. Some examples are:
- "Sold the bull with an *attitude*."
- "Breaker and switches are now labeled for electric."
- "Moved storage of formaldehyde to an area with better ventilation."
- "Provided a new pair of safety glasses to shop mechanic."
- "Maintenance of lagoon fence."
- "Shop (welding area) cleaned up and flammable materials removed from area.'
- "Many new fire extinguishers were added in hazard locations throughout the farm."

Workers' Compensation Insurance and Claims

Due to regulations in New York State, all of the 65 farms that participated in the AHAT project had workers' compensation insurance for their employees. Insurance premium rates are set by a rating board appointed by the state. In recent years, the premium payment which is based on total payroll, has ranged from $10-$18 per $100 payroll.

Sixty-one farms returned the "Workers' Compensation History Form"; the four farms that chose not to return the form did not receive an incentive payment for participating in the AHAT project because the incentive was a rebate based on the insurance premium paid. The farms were contacted several times regarding this element, but chose not to provide the information.

The total workers compensation premium paid during the fiscal year 1996 by the 61 farms reporting, was $634,070.72. The annual premiums paid ranged in amount from $625.31 to $58,484.00 with a mean of $10,394.60 and median of $4,592.30.

According to self-reporting by the 61 participating farm owners, 65 workers' compensation claims were filed in 1996. During the project evaluation period, each farm owner was contacted by an AHAT staff member and again asked if any workers' compensation claims were filed in 1997. Owners reported 47 claims filed during 1997. This indicates a 27% decrease in the number of claims filed during the project year.

It is interesting to note that of the 47 claims reported, 51% (24) of the injuries involved interactions with dairy cattle. Due to the trust that was established between farm owners and staff during the AHAT project, it is felt these self-reports were accurate reflections of the claims made.

Incentive: Rebate on Insurance Premium

The incentive rebate payment, as previously explained, was derived from a formula based on the major participation components of the AHAT project. Therefore, an individual farm owner's choice of what and/or how much to do determined the amount of the rebate received. It is important to note, once again, that one of the goals of this research was to observe what activities farmers would *voluntarily* complete to receive an incentive payment.

Of the 61 farms eligible for an incentive payment, 57% (35) completed 100% of the activities and their incentive equaled a rebate of 14% of their 1996 workers' compensation premium payment. The rebate for these farms ranged from one farm at $87.54 to five farms at $4,000.00, the maximum amount of rebate that an individual farm could receive which was established by the AHAT staff. Sixty-six percent (40) of the farms completed more than 90% of the activities (mean 92.5%). Only one farm had less than a 50% activity completion rate.

Evaluation of Farm Owner's Attitudes and Beliefs About Safety Training

Eighty-five percent (52) of the farm owners returned the retrospective "then/now" evaluation of attitudes and beliefs toward safety training that was mailed with their rebate check. Forty-eight of the surveys were complete and eligible for analysis. The purpose of this survey was to have the participants indicate if they felt a change in their attitudes had occurred while they were involved in the safety training portion of the AHAT project. The paired t-test results are shown in Table 5. There were statistically significant changes in farm owners' attitudes and beliefs concerning seven of the concepts presented. These changes indicated a positive shift in the owners' feelings about conducting safety training with their employees.

Table 5: Comparison of Farm Owners' Attitudes/Beliefs About Training Before and After Participation in AHAT. (n=48)

STATEMENT	Pre-AHAT (Mean)	Post-AHAT (Mean)	SD**	P*
1-Teaching adults is a lot like teaching kids.	3.021	2.979	1.336	0.830
2-Our employees will do as I instruct them to do regardless of how I do my own work.	2.312	2.021	1.398	0.155
3-It's hard to change employees' work practices because their habits are already in place.	3.375	2.708	1.191	**0.000***
4-Our employees are interested in improving their safety knowledge and skills.	3.188	4.021	1.018	**0.000***
5-I feel uncomfortable conducting safety training with our employees.	3.239	2.478	1.158	**0.000***
6-Our employees only pay attention to safety when they know I'm watching them.	3.000	2.333	0.953	**0.000***
7-Using an expert's advice rather than my personal experiences makes training more valid.	3.604	3.417	1.232	0.297
8-It's important for employees to be involved in choosing safety topics.	3.354	3.583	0.928	0.094
9-Safety training is much more important for our new employees than for our long-time employees.	2.894	2.234	1.147	**0.000***
10-Our employees will listen to a safety specialist more than to me.	3.396	3.188	1.071	0.184
11-Correcting an employee's mistake when it happens is an effective substitute for formal safety training.	2.896	2.188	1.071	**0.000***
12-I know how to be an effective trainer.	2.854	3.667	0.790	**0.000***

*$P < .05$. **SD - standard deviation

A second interesting comparison between attitudes about safety training in a control group surveyed at a major farm show and the before/after AHAT group responses is shown in Table 6. A two sample t-test was performed on the mean responses for the "before/control" groups and "after/control" groups.

There were several statistically significant differences in attitudes found between the participants in the AHAT project both before and after treatment and those attitudes of the general farm population that participated in the survey at the farm show. It appears that the control group had stronger agreement with the following statements than the AHAT participants either before or after the training period:

- Teaching adults is a lot like teaching kids.
- Our employees will do as I instruct them to do regardless of how I do my own work.
- It's hard to change employees' work practices because their habits are already in place.
- It's important for employees to be involved in choosing safety topics.

One final interesting result was that the control group appeared to have more confidence in their effectiveness as trainers than the AHAT participants did before they attended the train-the-trainer seminars and conducted employee training.

Table 6: Comparing Differences in Attitudes/Beliefs of AHAT Participants and Control Group (n=100)

Statement #	Before AHAT (Mean)	Control (Mean)	P*	After AHAT (Mean)	Control (Mean)	P*
1[a]	3.04	3.80	**0.003***	2.96	3.80	**0.002***
2[a]	2.33	3.09	**0.002***	2.00	3.09	**0.000***
3[a]	3.388	3.812	**0.022***	2.69	3.812	**0.000***
4[a]	3.204	3.89	**0.0001***	4.00	3.89	0.55
5[a]	3.26	2.39	**0.0005***	2.469	2.39	0.73
6[a]	3.00	2.67	0.13	2.327	2.67	0.12
7[a]	3.633	3.32	0.15	3.41	3.32	0.72
8[a]	3.388	4.06	**0.0004***	3.57	4.06	**0.017***
9[a]	2.85	2.54	0.24	2.21	2.54	0.21
10[a]	3.408	3.30	0.59	3.16	3.30	0.56
11[a]	2.94	2.84	0.69	2.16	2.84	**0.014***
12[a]	2.857	3.70	**0.000***	3.673	3.70	0.88

[a] Number corresponds with statement in Table 5. *$P < .05$.

Information was provided by the AHAT participants with regard to the number of hours of employee training they had conducted during the three months before AHAT. This same question was asked of the control group at the farm show. Fifty-seven of the AHAT participants reported conducting training for a range of 0 - 12 hours over the previous three months with a mean of 1.7 hours; the 50 control group members reported a range in training hours of 0 - 100 with a mean of 14 hours.

DISCUSSION

Hazard Abatement

Hazard abatement focused on guarding/shielding problems on tractors and equipment that were identified during a safety survey. The concept of improvement of work site safety through engineering controls was encouraged by mailing a list of farm specific guarding/shielding hazards to each owner. Participants could then choose which hazards they felt were most important to correct and consequently, which hazards they would correct. It is interesting to note that missing guards/shields had the highest rate of abatement (95%); 80% of PTO master shield hazards were corrected and 77.9% of PTO shaft shield hazards were abated. Missing or damaged SMV emblems were found in greater numbers, yet only 60.4% were corrected. Oddly, the cost for replacement for SMV signs is much lower than individual PTO master shields and shaft shields. It is believed that many of the missing master shields were still present on the farm, perhaps in the corner of the shop. Therefore, the replacement effort for the farm owner was minimal; the exact shield for that tractor was relatively easily located and installation was direct.

PTO shaft shields would not generally be laying around the farm so this category is more of a challenge to explain. Perhaps the belief that PTO entanglement is much more likely and more serious than roadway incidents (which may or may not be related to SMVs) lead participants to retrofit the PTO shafts. In addition, the owners' genuine concern for their employees may have influenced their decisions to retrofit PTO shaft shields over less perceived risks of roadway incidents.

The SMV emblems were replaced but at a lower rate than the other hazards. Was this because there were more SMVs needing replacement? Was it because people feel as though they are unimportant? The results would suggest that both of the above may be true in this sample. Hopefully, a new generation of SMV emblems with longer lasting materials and higher retro-reflective properties will allow the emblems to have a longer life span and therefore not require owners to replace so often.

Training Methods

Several different styles of training in a work place environment were demonstrated during the two training seminars. The details of three methods (formal, on-site, video) and their respective pros/cons were presented to the participants and sample training guides of each method were available in the *Farm Worker Training Manual*. There were no attempts made to persuade the participants to use a specific method for any particular topic. With this basis, it was very interesting to monitor the styles of training that owners/managers chose to use with their employees.

The vast majority (73%) of training sessions utilized the *on-site* method with the bulk of the remainder (21%) utilizing the *video-based* training. As was stated before, the *formal* trainings reported were off-farm meetings or guest speakers. These findings indicate that most of the owners were beginning to see themselves as educators and were taking the opportunity to perform training on the spot, at the point of importance or potential hazard. The *on-site* trainings allowed the participant-trainers to discuss familiar topics utilizing a more personal style versus a more traditional style of standing in the front of a classroom. The AHAT staff was aware that the initial perception of most adults regarding teaching is what they themselves experienced in high school and/or college - "the classroom setting". One of the goals of the seminars was to broaden the farm owners' views and use of training methods that might be better suited to the work environment.

Discussions of how to effectively "teach" safety to adults emphasizing the concept of using a 'team' approach were part of the seminars. These discussions focused on the characteristics of adult learners and differences between children and adults as learners.

Using videos effectively for training was demonstrated and participants were offered a six-segment video to use. Sixty-nine percent (47) took the video home as a resource. Those who used the video found that bringing employees into the owners home to view the video sometimes presented a barrier to learning. Employees were uncomfortable in the home; suggestions were made by other participants to use the break room or shop instead.

Training Subjects

The safety issues that farmers chose to provide training for were very interesting. Table 4 shows that 'animal handling' was the most frequently taught subject which, by the way, is also one of the most complex subjects to teach to employees. It encompasses many facets including understanding animal behavior, the relationship of the handler to and the overall attitude of the employee toward animals and engineering controls. Handling animals safely may require a whole new way of thinking for some individuals. It would seem that since many injuries occurring on farms are related to animals, the owners felt this was an extremely important topic. Perhaps the owners' new enthusiasm for education prompted them to attack issues in which they were hesitant to address in the past because they felt inadequate as teachers.

Tractor-related training was the second issue most frequently addressed. This reinforces the fact that dairy farm operations rely heavily on tractors for daily tasks. Most trainers used *on-site* training when addressing tractor operation topics. This allowed them to demonstrate and use hands-on methods for training. It also allowed them to observe employees in an effort to complete the "explain-demonstrate-try-observe-reinforce" education continuum.

Self-Reporting Accuracy

The design of this study relied heavily on building a trust or relationship with the participants and depending on their honesty in reporting their activities on the monthly reporting forms. This was a necessity since it would have been impossible for staff to accurately observe and record each farm's activities over the year-long project involvement which included a six month training period. Also, on-farm monitoring by AHAT staff might cause the farm employees to behave differently than their natural work habits.

Initially, participating farms had to supply their workers' compensation premium total and number of claims filed for the prior year (1996). Several farms shared paperwork on their 1996 claims during farm visits which proved the accuracy of self-reports. After hazard abatement activities on farms were self-reported, 25% of the farms were randomly visited to inspect these changes: over 95% of the corrections were found in the condition as reported. This reinforced the staff's confidence in the self-reporting system.

Farm Owners' Attitudes and Beliefs

Table 5 represents the statistical results of the evaluation of farm owners' attitudes and beliefs as measured in a retrospective pre/post test. Changes were noted with respect to specific attitudes about training and training employees. The following attitudes/beliefs showed statistically significant changes beginning with those in which the change (pre and post AHAT participation) was the greatest:

1) Employers realized to a greater degree that the employees are interested in improving their safety knowledge and skills.
2) Owners felt increased confidence in being effective trainers.
3) Owners were more comfortable conducting safety training for their employees after participating in AHAT.
4) Employers noted that correcting an employee's mistake when it happened was not a substitute for safety training.
5) Employers realized through the project that the employees not only pay attention to safety when they are being watched but also when not being openly observed.
6) Owners learned that the work habits of the employees are not "fixed in stone".
7) Owners concluded that safety training is important for old and new employees.

These important changes in attitudes and beliefs are vital in creating a proper mindset in employers toward establishing ongoing safety programs and were a measurable outcome of the AHAT project.

Empowerment and Attitude Change

The real heart of this project involved empowering farm owners and managers to make a positive difference in the safety of their operations. The role of the AHAT staff was to facilitate the owners' ability to recognize hazards, prioritize them for their businesses, and take action to address the problems. Key to the success of the project was teaching the farm owners about effective adult education and then allowing them to tailor that knowledge to their own farms. Who is better to train employees than their employer? After all, is it not ultimately in an owner's best interest to maintain a happy, productive, and safe workforce for his/her business? Would not the employees respect their own employer more than a hired consultant who comes in and tells them how to work?

The farm owners and managers who participated in AHAT did become empowered. Their attitudes changed regarding key training issues and, in turn, their ability to improve the knowledge and actions of their employees with respect to safety issues increased.

CONCLUSION

The Agricultural Hazard Abatement and Training (AHAT) project was immensely successful. Participating farmers voluntarily reduced hazards, began ongoing safety training programs, and observed safer work practices . Positive changes in the farm owners' attitudes about training were statistically significant. Additional work-site modifications were made in a tertiary level of positive effect on the farm. Most importantly, the injury rate among the 575 workers decreased during this project.

This program demonstrated that by using a positive incentive initially to get farms involved, farm owners can be empowered to improve their farm work-sites and influence the behaviors of their employees. Farmers like to be in control of their own destiny and are very capable of successfully managing a safe, profitable operation. We, as safety and health professionals, need to facilitate that learning process.

REFERENCES

Accident Facts, 1997 Edition. (1997). Itasca, IL: National Safety Council.

Aherin, R., Murphy, D., & Westaby, J. (1992). *Reducing Farm Injuries: Issues and Methods.* St. Joseph, MI: ASAE.

Bayer, D. (1984). The benefits of a farm safety program. *California Agriculture, 38*(1,2) 26-27.

Becker, W. & Shoup, W. (1986). *Cost/benefits of safety training for citrus workers.* Paper No. 86-1. Paper presented at the Summer Meeting of the National Institute for Farm Safety, Orlando, FL.

Denis, W. (1988). Causes of health and safety hazards in Canadian agriculture. *International Journal of Health Services, 18*(3), 419-435.

Goldenhar, L. & Shulte, P. (1994). Intervention research in occupational health and safety. *Journal of Medicine, 36*(7) 763-775.

Hallman, E., Pollock, J., Chamberlain, D., Abend, E., Stark, A., Hwang, S., May, J. (1997). *Tractor and Machinery Hazard Surveillance within the NYFFHHS Project.* Paper presented at the Summer Meeting of the National Institute for Farm Safety, Indianapolis, Indiana.

Meyers, J. (1997). *Injuries Among Farm Workers in the United States, 1993.* Cincinnati, OH: NIOSH.

Murphy, D. (1992). *Safety and Health for Production Agriculture.* St. Joseph, MI: ASAE

Poppendorf, W. & Donham, K. (1991). Agricultural hygiene. In Clayton, G.D. & Clayton, F.E. (Eds.) *Patty's Industrial Hygiene and Toxicology*, 4th ed. New York: Wiley Pub. Co.

New York Agricultural Statistics. (1997, November). *Farm Labor.* Agricultural Statistics Board, USDA (p.13).

Pollock, J. (1990). *Perspectives of New York Farm Safety.* Unpublished master's thesis. Cornell University, Ithaca, NY.

Rosenblatt, P. & Lasley, P. (1991). Perspective on farm accident statistics. *Journal of Rural Health, 7*(1) 51-61.

Sorensen, G., Stoddard, A., Ockene, J., Hunt, M., & Youngstrom, R. (1996). Worker participation in an integrated health promotion/health protection program: Results from the Wellworks project. *Health Education Quarterly. 23*(2) 199-203.

Stueland, D., Lee, B., & Layde, P. (1991). Surveillance of agricultural injuries in central Wisconsin: Epidemiologic characteristics. *Journal of Rural Health, 7*(1) 63-71.

Speaker Biographies (presented in alphabetical order)

Jennifer LaPorta Baker

Jennifer provides representation and counseling to agricultural employers on a wide range of labor and employment matters, including wage and hour issues under the Fair Labor Standards Act; the employment of minors; unemployment compensation; employment discrimination cases under Title VII of the Civil Rights Act of 1964; the Age Discrimination in Employment Act; the Americans with Disabilities Act; the Family and Medical Leave Act; the National Labor Relations Act; and other federal, state, and local statutes. She also advises agricultural employers on personnel matters including the preparation of personnel policies, employment applications, employee handbooks, regulatory compliance, and handling terminations and reductions in force. Jennifer contributes to journals and other publications on labor and employment issues and has presented several seminars on various employment law topics.

After graduating from Ithaca College in 1992, Jennifer attended The Catholic University, Columbus School of Law, in Washington, D.C. While in law school, Jennifer focused her studies on labor and employment law courses, and since her graduation in 1995, she has practiced labor and employment law exclusively. After practicing for two years in Pittsburgh, Pennsylvania, Jennifer joined McNees, Wallace, and Nurick in Harrisburg, Pennsylvania, where she is currently an associate in the Labor and Employment Practice Group. McNees, Wallace, and Nurick represents agricultural employers throughout Pennsylvania and the Eastern United States.

This speaker's paper
"Hiring with and without a Contract"
begins on page 45

John C. Becker

John Becker is a professor of agricultural economics on the faculty of Penn State University, University Park. He is a 1969 graduate of LaSalle University with a B.A. degree in economics and a 1972 graduate of the Dickinson School of Law with a J.D. degree. He is a member of the Pennsylvania Bar and is of Counsel to the Camp Hill, Pennsylvania, law firm of Zeigler and Zimmerman, P.C. His teaching programs and publications include bulletins, circulars, and independent study courses that focus on legal issues such as environmental law and regulation, estate tax, estate transfer, land owner liability, real estate tax assessment, and employer-employee issues. He is Director of Research at the Agricultural Law Research and Education Center of the Dickinson School of Law at Penn State University and an adjunct faculty member.

Mr. Becker is a member of the Centre County, Pennsylvania, and American Bar Associations and the American Agricultural Law Association. He is a past chairperson of the Agricultural Law Committee of the Pennsylvania Bar and has served as a director of the American Agricultural Law Association and vice-chair of the Agricultural Law Committee of the General Practice Section of the American Bar Association. In 1994 he was elected a fellow of the American Bar Foundation.

Mr. Becker has organized and presented several continuing education programs for the Pennsylvania Bar Institute, Dickinson School of Law, the American Agricultural Law Association, and the American Association for the Advancement of Science. His published legal research appears in *Drake Law Review, Dickinson Law Review, Indiana Law Review, William Mitchell Law Review, the Drake Journal of Agricultural Law, The Journal of Soil and Water Conservation,* and the *Pennsylvania CPA Journal.*

Mr. Becker is a retired member of the Pennsylvania Army National Guard, having served as Command Judge Advocate of the 28th Infantry Division (Mech.) holding the rank of Colonel in the Judge Advocate General Corps.

*This speaker's paper
"Elements of an Employment Contract"
begins on page 53*

James G. Beierlein

Jim Beierlein (pronounced buyer-line) is a professor at Penn State's University Park Campus. Since 1977 he has taught agribusiness management both on and off campus. He is the author of seven books on agribusiness management and marketing. Jim was a manager and later general manager of a twenty-three-outlet restaurant that employed more than 1,000 people. In that capacity he received two awards for his management.

Since 1987, he has conducted management programs for agribusinesses, called *Managing for Success*. These programs cover topics such as financial management, marketing and advertising, time management, strategic planning, and people management. He is a recent winner of his college's outstanding teaching award.

Dr. Beierlein is a native of New Jersey. He is a graduate of Rutgers and Purdue Universities. He and his wife are "enjoying" the teenage years with their two children.

This speaker's paper
"Getting the Most from Your Employees"
begins on page 35

Bernard L. Erven

Bernie Erven is professor of agricultural economics and extension specialist in the Department of Agricultural Economics at The Ohio State University. His teaching, extension and research activities are in human resource management, farm management, and business management. He teaches courses in agribusiness management, human resource management in small businesses, and career management.

Bernie's extension program focuses on labor management topics such as hiring, training, motivation, compensation, and performance evaluation. He also does workshops on communication techniques, handling the stress of management, time management, management succession, and human relations in family businesses.

He has degrees from The Ohio State University and the University of Wisconsin.

He has been a visiting professor at Cornell University. Bernie's foreign assignments include seven months teaching at a school for refugee boys at Jericho, Jordan, and 3 1/2 years university teaching in Brazil.

Bernie has twice received the Ohio State University Award for Distinguished teaching. He has also received the American Agricultural Economics Association Teaching Award, the Ohio State University Gamma Sigma Delta Extension and Teaching Awards, and the American Agricultural Economics Association Group Extension Award.

This speaker's first paper
"How Much Are Your Employees Worth?"
begins on page 1

His second paper
"Recruiting and Hiring Outstanding Staff"
begins on page 24

Al French

Al French is the Coordinator of Agricultural Labor Affairs in the Office of the Chief Economist of the U.S. Department of Agriculture. He is USDA's lead man for agricultural labor policy issues. He was responsible for drafting the regulations defining "seasonal agricultural services," which determined the types of workers eligible for legalization under the Special Agricultural Worker (SAW) program of the Immigration Reform and Control Act of 1986. He was also USDA's consultant to the Department of Labor in drafting the Temporary Agricultural Worker (H-2A) program regulations. Al consulted with the Environmental Protection Agency in the promulgation of its agricultural Worker Protection Standards.

Al was formerly the State Executive Director of USDA's Agricultural Stabilization and Conservation Service (now Farm Service Agency) in Florida.

Prior to government service, Al was Director of Public Affairs for the Florida Farm Bureau. He brought to that job 15 years of experience as a consultant in agricultural labor relations with the Management Research Institute.

A native of Florida, he grew up in his family's citrus groves. After graduating from Palm Beach High School, he attended Washington and Jefferson College in Pennsylvania and Tulane University in New Orleans, Louisiana.

Al and his wife, Beverley, have four children and currently reside in Washington, D.C.

This speaker's first paper
"Guest Workers in Agriculture: The H-2A Temporary Agricultural Worker Program"
begins on page 58

His second paper
"Farm Employment Rules and Regulations: What You Need to Know"
begins on page 72

Lisa A. Holden

Lisa received a B.S. degree in animal bioscience from Penn State in 1988 and an M.S. degree in animal science from the University of Maryland in 1990. In 1992, Lisa began working with the Dairy Management And Profitability (Dairy-MAP) program, and in 1993 she completed her Ph.D. in animal science at Penn State.

Since 1994, Lisa has been working as an assistant professor in the Department of Dairy and Animal Science at Penn State. She has responsibilities in both extension education and research in the areas of dairy management and dairy nutrition. In addition to directing the Dairy-MAP program, Lisa is involved with dairy advisory teams and works with both dairy producers and agribusiness representatives.

*This speaker's paper
"Communicating the Mission to All Personnel"
begins on page 10*

Robert A. Milligan

Robert A. Milligan received his B.S. and M.S. from Michigan State University and a Ph.D. from the University of California at Davis. Bob is an agricultural economist at Cornell University specializing in small business management. Bob's professional interests include management education, human resource management, and organizational development for small-business managers. Professor Milligan teaches an undergraduate class entitled "Managing Human Resources in Small Businesses" and presents thirty to forty workshops a year for small-business managers. Bob is a coauthor of a book on personnel management for the Golf Course Superintendents of America and is a member of their teaching faculty.

Bob received the Gamma Sigma Delta Certificate of Merit for extension activities in 1992; a commendation recognizing his pioneering efforts in teaching management concepts in 1995; and the Award of Recognition from the NYS Association of Agricultural Agents in 1996. He was named a Clark Professor of Entrepreneurship and Personal Enterprise in 1998.

This speaker's first paper
"How Much Are Your Employees Worth?"
begins on page 1

His second paper
"Leadership: Coaching to Develop People"
begins on page 40

Walter C. Montross

Walter is the Golf Course Superintendent at Westwood Country Club in Vienna, Virginia. This is his thirtieth year in the business, the last twenty of which he has been a GCS.

Walter has been a Certified Golf Course Superintendent since 1984. He graduated from the University of Maryland in 1975 with a degree in turf and golf course management.

Walter has also been very active at both the local and national levels of the Superintendents Associations. He is serving as the 1999 president of the Mid-Atlantic GCS, and he held the same position in 1989. He is a past president of the Greater Washington GCS and the University of Maryland, Institute of Applied Agriculture Alumni Association. He is now serving on the Communications Committee for the Golf Course Superintendents Association of America and previously served on the Career Development Committee.

This speaker's paper
"Managing the Multicultural Workforce"
begins on page 14

Michael D. Pipa

Mr. Pipa is a civil trial lawyer with the law firm of Mette, Evans, and Woodside in Harrisburg, Pennsylvania. He is a native of Northumberland County, Pennsylvania, and practices in the State and Federal Courts in the Central Pennsylvania area. His practice includes concentrations in employment discrimination, personal injury, and product liability law.

Mr. Pipa received an engineering degree, with honors, from Lafayette College and served as a market specialist with the General Electric Company before attending law school at the George Washington University in Washington, D.C. He received his law degree, with honors, in 1988, and was a member of the George Washington Law Review. He practiced with the national firm of McKenna and Cuneo in Washington, D.C., before returning to his home state to join Mette, Evans, and Woodside in 1989. Mr. Pipa was recognized in the Ninth Edition of *Who's Who in American Law*.

This speaker's paper
"Discrimination in the Workplace"
begins on page 65

Don R. Rogers

Don Rogers serves as a senior farm business consultant with First Pioneer Farm Credit, covering the territory of New Jersey, New York, and Southern New England. He heads Farm Credit's consulting group that provides fee-based financial services to farmers and agribusiness. His specialties include farm management with emphasis on financial and labor issues, business expansion, and family transfers.

Don spent 18 years with Farm Credit Banks of Springfield developing consulting services for the Northeast. Since he came to Farm Credit in 1977, his consulting work has been with over 2,000 farm families. He also serves as a teaching resource for Farm Credit's staff and agribusiness in the area of farm management. He has conducted workshops and presentations to numerous farm groups on management topics including business planning, farm transfers, and family communications.

In 1997, Don received designation as an Accredited Agricultural Consultant (AAC) from the American Society of Farm Managers and Rural Appraisers. Having been active on its Consulting Committee for the past 8 years, he is one of the first class to receive accreditation.

Don started his long agricultural career with New York State Cooperative Extension in Columbia and Dutchess Counties as an agricultural agent — specializing in farm management. A Cornell agricultural graduate, he is married and has four children. Don grew up on a small farm outside Cooperstown, New York. He presently resides in Longmeadow, Massachusetts.

This speaker's paper
"Performance Feedback"
begins on page 19

David M. Scott

Dave was raised on the family dairy farm in Butler County, Pennsylvania, until he was eleven years old, when the Commonwealth purchased it as part of a new state park. He continued to be involved in agriculture until graduating from East Stroudsburg University with a Bachelor of Science degree in biology in 1975.

Dave entered state service with the Pennsylvania Department of Agriculture in 1978 with the Bureau of Animal Industry as a livestock disease control technician working with the livestock, dairy, and poultry industries in the eradication and control of tuberculosis, brucellosis, pseudorabies, rabies, and the avian influenza. Dave transferred to the Bureau of Plant Industry in 1989 and served nine years as an agronomic products inspector. These duties included inspection sampling and enforcement activity for feed (commissioned FDA Inspector), fertilizer, lime, seed, and pesticides. The pesticide activities entailed responsibility of enforcing both state and federal statutes. In 1997, Dave was honored for his work by EPA and named "The Outstanding Inspector of the Year" for EPA Region III. Dave was selected in September 1998 by the Pennsylvania Department of Agriculture to serve as the pesticide certification and education specialist, responsible for programs to develop and maintain: pesticide competency exams, continuing education programs, and licensure of pesticide businesses, dealers, and applicators across the Commonwealth.

This speaker's paper
"EPA Worker Protection Standard (CFR Title 40, Part 170)"
begins on page 92

Conference Notes

Conference Notes

Conference Notes